FRANKENSTEIN URBANISM

This book tells the story of visionary urban experiments, shedding light on the theories that preceded their development and on the monsters that followed and might be the end of our cities. The narrative is threefold and delves first into the *eco-city*, second the *smart city* and third the *autonomous city* intended as a place where existing smart technologies are evolving into artificial intelligences that are taking the management of the city out of the hands of humans.

The book empirically explores Masdar City in Abu Dhabi and Hong Kong to provide a critical analysis of *eco* and *smart* city experiments and their sustainability, and it draws on numerous real-life examples to illustrate the rise of urban artificial intelligences across different geographical spaces and scales. Theoretically, the book traverses philosophy, urban studies and planning theory to explain the passage from eco and smart cities to the autonomous city, and to reflect on the meaning and purpose of cities in a time when human and non-biological intelligences are irreversibly colliding in the built environment.

Iconoclastic and prophetic, *Frankenstein Urbanism* is both an examination of the evolution of urban experimentation through the lens of Mary Shelley's *Frankenstein*, and a warning about an urbanism whose product resembles Frankenstein's monster: a fragmented entity which escapes human control and human understanding. Academics, students and practitioners will find in this book the knowledge that is necessary to comprehend and engage with the many urban experiments that are now alive, ready to leave the laboratory and enter our cities.

Federico Cugurullo is Assistant Professor in Smart and Sustainable Urbanism at Trinity College Dublin, Ireland.

FRANKENSTEIN URBANISM

Eco, Smart and Autonomous Cities, Artificial Intelligence and the End of the City

Federico Cugurullo

Routledge
Taylor & Francis Group

LONDON AND NEW YORK

First published 2021
by Routledge
2 Park Square, Milton Park, Abingdon, Oxon OX14 4RN

and by Routledge
52 Vanderbilt Avenue, New York, NY 10017

Routledge is an imprint of the Taylor & Francis Group, an informa business

British Library Cataloguing-in-Publication Data
A catalogue record for this book is available from the British Library

Library of Congress Cataloging-in-Publication Data
Names: Cugurullo, Federico, author.
Title: Frankenstein urbanism : eco, smart and autonomous cities, artificial intelligence and the end of the city / Federico Cugurullo.
Description: Abingdon, Oxon ; New York, NY : Routledge, 2021. | Includes bibliographical references and index.
Identifiers: LCCN 2020052989 (print) | LCCN 2020052990 (ebook) | ISBN 9781138101760 (hardback) | ISBN 9781138101784 (paperback) | ISBN 9781315652627 (ebook)
Subjects: LCSH: Urbanization--Philosophy. | Sustainable urban development. | Smart cities. | Artificial intelligence--Social aspects.
Classification: LCC HT153 .C844 2021 (print) | LCC HT153 (ebook) | DDC 307.7601--dc23
LC record available at https://lccn.loc.gov/2020052989
LC ebook record available at https://lccn.loc.gov/2020052990

ISBN: 978-1-138-10176-0 (hbk)
ISBN: 978-1-138-10178-4 (pbk)
ISBN: 978-1-315-65262-7 (ebk)

Typeset in Bembo
by Taylor & Francis Books

For my parents who taught me to fear no monster

CONTENTS

FIGURES

ACKNOWLEDGEMENTS

On an unusually rainy day, many years ago, my mum took me to see Branagh's *Frankenstein*. The cinema was old and dark, and it felt to me like we were in a different world, far away from my hometown Cagliari called *the city of sun* where the sky is rarely grey. Maybe I did enter another world, because after watching that movie, and later reading the book that had inspired it, my life was never the same. Thank you mum for buying that ticket and holding my hand.

Since then, it has been an incredible journey and quest for knowledge culminating now in this book. Thank you dad for cultivating my passion for travelling. Thank you Ele, my sister, for sparking my interest in urban design (and sorry for missing your graduation when I was immersed in the fieldwork in the Middle East). Thank you aunt Lalla, uncle Mau and aunt Vale for sharing your inspiring travel stories. And thank you Salvatore, Wanda and Angela, the best grandparents I could possibly have, who unfortunately saw only the beginning of my journey.

My obsession with ideal cities was transmitted by Professor Gian Giacomo Ortu when I was a master's student at Cagliari University. Thank you Prof. My intellectual debt is enormous.

Later when I started my PhD at King's College London, I was extremely lucky to be supervised by Professor Rob Imrie and Dr Clare Herrick. Thank you both for pointing me in the right direction and, above all, for being patient when I was stubbornly going the opposite way. Eventually, I ended up in Masdar City and it kind of worked out. Thank you Nick, Sam, Barbara, Rebecca, Marta, Hyungguen and Maria, my then PhD fellows, for miraculously managing to make me feel at home in a crazy city like London.

The idea to write this book was cultivated during my time as a lecturer at the University of Manchester, where it was impossible for me to draw a line between colleagues and friends. Thank you Ali, Bill, Joanne, Sergio, Filippo, Lucas, JZ,

Martin, Stefan, Saska, Kevin, Noel, Sarah, Jamie, Jen, Jonny, Helen, Erik, Maria and Saskia for making every day at the office a beautiful memory.

Ideas became words in Trinity College Dublin. Thank you Cian, Phil, Padraig and Anna for bravely going through the book's initial wild drafts, and for keeping the door of your office open when I needed your opinion. On a related note, my gratitude goes to Andy Karvonen, Paddy Bresnihan, Rob Cowley and Pauline McGuirk who, like sapient alchemists, carefully checked the narrative formula of the book, and to my editors Andrew Mould, Egle Zigaite and Faye Leerink for believing that this unorthodox formula could work.

In order to craft the book as it is now, I had to travel around the world first. I certainly would have got lost if it had not been for the friends who shared the road with me. Thank you Gianfranco, Berna, Fehmina and Antonio (Dubai), Suraiya (Sharjah), Nate (Hong Kong), Mona (Aswan), Wen (Taipei), Ioanna (Athens), Cath, Simon, Elisabeth and Sofia (Baltimore), Andrea, Marina, Jacob and David (Freiburg), Stefano and Eleonora (Torino), Martina, Roman, Andrea, Dorothea, Carol, Ransford, Micha and Weimu (Dublin), Giachi, Ste, Davi, Selly, Matte, Fede, Simo, Gigi, Lorenzo and Elisabetta (Cagliari).

Thank *you* for choosing to read this book and to my many students in Dublin and Manchester who, unlike you, had no choice but to attend my lectures and hear me raving about *Frankenstein* while they just wanted to know how cities can become sustainable.

And finally, thank you Mary Shelley for filling my mind with splendid nightmares, and Lingli for filling it with dreams.

1

PROLOGUE

Of cities and monsters

Introduction

The narrative of this book follows Mary Shelley's novel *Frankenstein*. Stripped down of their technicalities and embellishments, the two books are essentially about the perennial tension between ideas, theories and visions on one side, and facts, practises and results on the other. They both focus on how what is believed to be unsustainable can, in theory, become sustainable, and on the monsters that the reckless pursuit of a development ideal can in practise generate. In Mary Shelley's book, the protagonist is a doctor named Victor Frankenstein. Victor strongly believes that the human being is not sustainable. Humans, he argues, are fragile creatures. They are vulnerable and prone to diseases. The human being was born to die. Whether through illness, injury or simply ageing, the human body will eventually decay and, for this reason, the human condition is one of unsustainability. These adamant beliefs which Victor keeps inside, are rooted in what is a tragic past. His mother died of scarlet fever when he was 17: an event which pushed him to embark on a quest for the creation of the ideal and perfect being, immune to death and to all the calamities in the world.

While Mary Shelley's novel explores the *human equation* or, in other words, the formula for the creation and enhancement of the human being, the present book deals with the *urban equation*. A plethora of cities are showing evidence of unsustainability. As first pointed out by Aristotle in ancient Greece, the city was originally created by humans to support human life. However, many cities seem to have now become dangerous places where people die prematurely and live a life of misery. Numerous academics, policy-makers, architects and planners have, like Victor Frankenstein, dark feelings about the past and the present, but see hope in the future. There is a growing awareness of the unsustainability of cities, followed by the realization that a formula for urban sustainability *has* to be found, and by the conviction (or perhaps the illusion) that such formula *can* be found. This mysterious formula is what the book calls the urban equation. It is a method meant to identify the core elements of a city which combined in a given proportion are supposed to produce sustainable cities. How urban equations are being formulated nowadays, and the extent to which they are actually capable of achieving urban sustainability, will be recurring points of critical analysis throughout the book.

The words of Mary Shelley tell the story of a desperate scientist with the soul of a philosopher, who seeks to reshape the human fabric in search of perfection but, instead, ends up creating a monster. It is a story of intellectual inquiry, experiments and tragic revelations. This book tells the story of urban experiments. It is a story of visionary urban projects, of the theories that preceded their development and of the monsters that followed, and might radically alter cities to the point of ending them. The narrative is threefold and delves first into the *eco-city*, second the *smart city* and third the *autonomous city* intended as a place where existing smart technologies are evolving into artificial intelligences which are taking the management of the city out of the hands of humans. On these terms, *Frankenstein urbanism* means both a way of narrating the evolution of urban experimentation from *eco-* and *smart-*city experiments to *autonomous* cities, by using Shelley's *Frankenstein* as a framework, and an urbanism whose product resembles Frankenstein's monster: a fragmented entity which escapes human control and human understanding.

The urban equation

Victor Frankenstein's quest is ambitious not simply because its goal, creating the perfect human being, is enormously challenging from a scientific perspective. Victor's quest deals with nebulous concepts and questions which, despite many attempts, have never been fully answered. What the source of human life is and, above all, what the essence of being human is, are questions which have been puzzling scientists and philosophers alike for millennia (Scruton, 2017). It is therefore incredibly difficult, if not impossible, to develop the formula for the ideal human being, when the meaning of *human* is unclear in the first place. One might say that Victor has already failed even before embarking on his quest. The experiment attempted by the young scientist is flawed due to a lack of conceptual clarity. Victor cannot create what he does not fully understand.

The same problem can be seen in the context of urban sustainability. Creating a formula for sustainable urban spaces is not simply a scientific endeavour. Like the concept of being *human*, the notion of *urban* escapes a universal definition and its understanding is often ambiguous. The meanings of being urban, being a city and, more generally, urbanization, are not self-evident and not necessarily interchangeable. Rather than being solid like the stones of a city, the urban is a labyrinth made of mist, whose exploration can lead anywhere and nowhere. The term *urban* comes from the Latin word *urbanus* meaning *belonging to* or *of* the *urbe* which in turn means *city*. However, as repeatedly pointed out in the field of urban theory, what constitutes a city is hard to define, and this is where the academic debate tends to begin (Harding and Blokland, 2014; Iossifova et al., 2018; Jayne and Ward, 2016;

McNeill, 2016). Here the philosophy of Aristotle can be both a door to enter the labyrinthine debate over the urban equation, and a lantern to shed light on it. The remainder of this section draws upon Aristotelian philosophy to identify and discuss the core dimensions of urban settlements. Aristotle's thinking will be combined with insights from contemporary urban theory not to provide an absolute definition of the terms *city, urban* and *urbanization*, but to clarify how these notions are interpreted specifically in this book in relation to the study of eco-cities, smart cities and cities populated by artificial intelligences. In so doing, the chapter introduces a series of fundamental concepts which will be examined in more detail later in the book.

For Aristotle (2000), the city is the *ultimate* form of a *human community*. The emphasis on the words 'ultimate' and 'human community' is meant to highlight two key aspects that constitute a city. First, the city is not the only typology of human community. Being the ultimate one implies that a city is part of a process of development and, as such, it does not appear out of nothing: it comes from something. From an Aristotelian perspective, in order to understand where the city comes from, the focus is directed towards the evolution of human communities. For the Greek philosopher, a community (*koinonia*) has three forms. Each one representing a stage of human development. The first form is the family; the second is the village which unites different families; the third is the city which brings together different villages. It is important to note that underpinning the philosophy of Aristotle is the concept of *teleology*: the idea that everything has an inner potential or a final cause which can be reached through a process of development (Aristotle, 1996). What constitutes a city, therefore, does not manifest itself only in the city. The seeds of the urban are in the family living an isolated life. Urban seeds then grow into a small village. It is only through the evolution of families and villages that the urban flourishes and becomes evident in the city.

In his studies, Aristotle discusses the qualities that characterize human communities, including cities (Aristotle, 2000, 2004). According to his philosophy, the city is not only physical, and it goes beyond the built environment: it is also social and political. To explore this key point in-depth, semantic clarity needs to come before conceptual clarity. Aristotle uses three distinct terms in *The Politics* whose wording in ancient Greek can be found in copies of the classical text (see, for instance, Dreizehnter, 1970; Ross, 1957). These three words are *oikos* (οἶκος), *core* (κώμη) and *polis* (πόλις), and they represent the threefold evolution of human communities from families to villages and ultimately to cities. The oikos has three interconnected meanings. First, as a form of koinonia (community) it has a social meaning. In this sense, the term oikos signifies *family*. As mentioned above, for Aristotle, the family is the basic and earliest form of community or, in other words, social organization. The oikos, however, in the standard society of ancient Greece, does not comprise only two parents and their children, but also the slaves and the animals which serve the family. For this reason, translators like Sinclair and Lord prefer to use the word *household*, instead of family (see Aristotle, 1992, 2013). Second, the oikos has a political meaning, since the household requires a form of government in order to function. For Aristotle, the different relationships within the household are regulated in a patriarchal way. The oikos is run by the husband and then, from a hierarchical point of view, comes the son, the wife, the slave and the animal. At the time of Aristotle, there were of course other typologies of domestic governance which prove the complexity and necessity of the political dimension of the oikos. In Sparta, for example, women had a stronger control over the household since men were often at war, away from home, or living in barracks (Blundell, 1995). Third, the household has a physical dimension and, as such, it has a physical location and a physical shape or, in architectural terms, a *built environment*. On these terms, oikos can be translated as *house*. Before the formation of villages and cities, the oikos extended beyond the house intended as an independent building, and included pieces of land used by the family for agriculture (Carr Rider, 2014; Morachiello, 2004). It was therefore an extended environment comprising the spaces necessary to obtain the resources needed by the family.

These three qualities (the social, the political and the physical) repeat themselves, like DNA strands, in the remaining categories of human community identified by Aristotle, *core* and *polis*, but they become more complex in terms of size and organization. The village, as the sum of different households comprises members of different families and, thus, requires a more sophisticated type of political organization, in order to function in a harmonious way. It also requires more space, a larger built environment, as well as more resources. The polis (whose matrix is the oikos) shares the same characteristics of the household and the village. The polis is first a social entity, since it unites numerous people originally from different villages or born in the city itself. Second, the polis is political because of the government that is required to coordinate all the activities that underpin its life and economy. Finally, it has a tangible physical quality, due to the many buildings, infrastructures and vast territory that its population needs to prosper (Aristotle, 2000). For these physical and socio-political dimensions, in the literature the term polis has been translated as both *city* and *state*. The specific lexicological choice varies from translator to translator. Laurenti, for example, suggests using *state* when Aristotle discusses the political organization of the polis, and *city* when the subject of the discussion is its physical structure (see Aristotle, 2005). For the purpose of this study, what matters is the understanding of the city as simultaneously and intrinsically social, political and physical. This view is in sync with twenty-first-century urban theory in which the nature of cities is approached in a multi-dimensional way, and the city is seen not merely as a physical artefact, but rather as a hyper complex entity made of social, political, economic and cultural processes (Heynen et al., 2005). This is a line of thought which will cut across the whole book, in order to shed light on the multiple dimensions of eco-city projects, smart-city initiatives and autonomous cities.

In Aristotelian terms, as discussed above, cities are part of a process of evolution. In this sense, urbanization can be understood as a socio-political and physical process whereby households become villages and villages turn into cities. This Aristotelian perspective should not be interpreted in a normative way given that there are always exceptions to take into account, such as new cities built from scratch which are not extending from the family. However, this perspective is useful for seeing the city as part of a process of development which, in turn, leads to two important considerations. First, the scale of urban development tends to be addressed in contemporary debates in urban geography in a non-centric way, to avoid picturing the city as a rigid urban unit and as the sole manifestation of the urban. The work of Brenner (2019) and Schmid (2018), for instance, shifts the understanding of the urban outside the city and its centre, looking at the different scales through which the urban manifests itself. On these terms, as posited by Aristotle, the urban, although more prominent and evident in the city, can be found not only in the city. From this point of view, urbanity also lies, to a lesser degree, in the small settlements outside the boundaries of a city and, more generally, in all the infrastructure, supply chains and socio-political and economic activities that pivot around cities. The question of scale relates to the question of size: the amount of physical space that the built environment covers and where cities' activities extend to. For Aristotle, a household unit is relatively small when compared to a village, and a village is relatively small when compared to a city. Urban development is therefore seen as a process of growth through which urban settlements, and related infrastructures, supply chains and activities become bigger, thus covering and influencing more geographical spaces. In the fourth century BC, Aristotle could not predict that some cities were going to become megacities. He could also not foresee that urban spaces would exponentially grow and multiply, creating a condition that a strand of urban studies defines as *planetary urbanization* in which the boundaries of cities are becoming blurred (Brenner, 2014; Lefebvre, 2014/1987; Peake et al., 2018; Williams et al., 2020; Wilson and Jonas, 2018). Urban spaces, infrastructures and services are spreading globally, thereby transcending the city as a self-contained unit. The world is opening up as the stage where the urban spectacle takes place.

Second, seeing the urban as a process, through the lens of Aristotle's philosophy, points towards what sustains the process. For Aristotle (2000), urban growth necessitates a number of resources. The development of villages, their growth and integration into a city, and the genesis and development of the city itself, require food, metals, stones, animals and, of course, humans. In the first part of *The Politics*, the Greek philosopher lists some of the key arts that are needed in order to cultivate the seeds of the urban. Arts which range from agriculture to metallurgy, and from fishing to trade. It is again a question of scale. He notes, for instance, that 'in the first form of koinonia, which is the family, it is obvious that there is no purpose to be served by the art of exchange. Such purpose emerges only when the community is larger' (Aristotle, 2000: 25). Ultimately, Aristotle sees a human community, across its three main incarnations, as a living entity which consumes resources to grow, and the city, being the largest typology of human community, as the entity whose life requires resources the most. This perspective reflects current studies on urban metabolism, which put emphasis on the many flows of materials and energy underpinning the life-cycle of cities (Beloin-Saint-Pierre et al., 2017; Conke and Ferreira, 2015; Pincetl et al., 2012).

It is also important to note that urban development is not geographically homogenous, meaning that the three forms of community identified by Aristotle can be present, at the same time in different geographical spaces. In other words, the village does not always and everywhere replace the household and, similarly, the city does not absorb all villages. Although the city is his ideal form of community, Aristotle is the first person to recognize that cities can be found next to villages around which single households coexist. What Aristotle insists on is the urban nature of the human being which, he argues, should not live outside cities. Out of the city are only immortal gods which are born perfect and, as such, can live independently, while due to their limitations, out-of-the-urban humans would be vulnerable and 'in the position of a solitary advanced piece in a game of draughts' (Aristotle, 2000: 10). For the Greek philosopher, humans are essentially urban animals which need urban settlements to survive and cities, in particular, to develop and fulfil their potential. From a theoretical perspective this is an important point, inasmuch as it captures the idea that the condition of humanity depends on the condition of urbanity and that, for humankind to flourish, cities are necessary. Regardless of what philosophical stance is taken into account, from a practical point of view it is a well-known fact that most of the global population now lives in cities, and human life has largely become urban life. These reasons alone would suffice to make the urban equation an unavoidable research topic. If human nature, and therefore also the urban nature of the global population's distribution, cannot be changed, it is then imperative to understand how cities can become more sustainable. In order to do so, it is crucial to first unpack the meaning of urban sustainability.

Urban sustainability

As this book will show, a single formula for urban sustainability does not exist. Despite the influence of global discourses, the very idea of sustainability tends to vary according to specific geographical contexts, under the influence of local cultures, politics, economies and physical environments (Whitehead, 2003, 2007). Likewise, the notion of what makes an urban space sustainable is geographically sensitive (Angelo and Wachsmuth, 2020; Hansen and Coenen, 2015; Truffer et al., 2015). Ultimately, as Lefebvre (2009: 31) remarks, every society 'produces a space, its own space' typical of its socio-cultural attributes, and it would be therefore problematic to theorize a universal sustainable urbanism. Yet, it is possible to recognize some key aspects of urban sustainability, in line with the key aspects of cities, which are common across different geographical spaces. To this end, the philosophy of Aristotle can be used again as an entry point to navigate the complex debate over the meaning and practise of sustainable urbanism. The aim here is not to provide an in-depth analysis of the concept of urban sustainability, but rather to shed light on its complexity and identify its core dimensions: an exercise that will be related later in the book to specific case studies and their formulas for a sustainable urban development.

First, there is the issue of the scale of urban development. For Aristotle, as noted before, the city is the ideal form of human community and in *The Politics* he stresses the importance of carefully defining its size and boundaries. From a political point of view, for him, a large city can be dysfunctional, inasmuch as a good form of government must be based on the citizens' knowledge of each other. Within the ideal political system of Aristotle, citizens can vote for the election of their governors, thereby determining who will rule the city and make important decisions which will eventually affect the life of every individual. However, 'both in order to give decisions in matters of disputed rights, and to distribute the offices of government according to the merit of candidates, the citizens of a city must know one another's characters' (Aristotle, 2000: 262). In cities where the population is too large, Aristotle argues, it becomes almost impossible for their inhabitants to even see each other, let alone know each other intimately. In these cases, government and elections 'operate by guesswork' (Aristotle, 2000: 263). In terms of geography, urban planning, urban design and economics, the Greek philosopher points out that the larger a city is, the harder it is to find space for it, to construct enough buildings and infrastructures, and to obtain and circulate all the resources that its population requires, particularly 'commodities which the city does not itself produce' (Aristotle, 2000: 265).

From an urban sustainability perspective, the problem of scale is today a pressing one (Bettencourt and West, 2010; West, 2017). The reasons are not dissimilar from those highlighted by Aristotle. The governance of large cities, particularly when it comes to the development of urban policies targeting sustainability, is weighted down by the myriad of actors, offices and procedures that frequently populate overgrown political entities (Bulkeley et al., 2014a; Rode et al., 2020; Vitz, 2018). The materiality of large-scale cities is per se problematic, even in politically efficient contexts. On the one hand, urbanization is commonly understood as *production of space* (see Lefebvre, 2009). However, on the other hand, the formation of space, whether in the shape of housing, industry or transport, for instance, requires the *destruction of space*. The process of urban development does not take place on a blank canvas, but rather on complex and, in many cases, fragile ecosystems, such as rivers, forests and lakes. Urbanization creates space, while simultaneously destroying the space that was previously there. In ecological terms, a greater scale of the urban often implies a greater environmental degradation (Chen et al., 2020). Emblematic is the case of China where the physical growth of cities has caused, since the 1990s, major loss of natural habitat. In 2011, the urban population of China (which was barely 26 per cent in 1990) reached 51.3 per cent, through an incessant production of built environments which wiped out approximately a quarter of all the country's forest and water coverage (He et al., 2014; Li et al., 2015; Wang et al., 2015; Zhang et al., 2020). Being ecologically sensitive is thus key to the formulation of a sustainable urbanism, and this is a theme which will become prominent later in the book when the narrative will centre on *eco-cities* supposed to combine the science of ecology with the art of city-making.

Urban history shows that this is not a new phenomenon. As Mumford (1961) notes, the environmental impact of urbanization became prominent and evident in the sixteenth century, particularly in Europe, when the production of urban spaces began to mean a drastic reshaping of the physical geography of a region. Mumford (ibid.) points out that while the medieval city had grown on and around surface features, the baroque city was less sensitive to local topography, and tended to impose its regular layout on the environment. Land levelling is a trademark of sixteenth-century urbanism: a practise which was implemented to remove irregular areas, and build straight streets essential for the development of early forms of trade-based capitalist economies (Conforti, 2005). What is unprecedented and worrying nowadays is the scale of the phenomenon. Although with notable geographical differences, the planet's urban mass is growing exponentially (Elmqvist et al., 2018; Melchiorri et al., 2018). While the population of European urban settlements is shrinking, in Asia, Africa and Latin America existing cities are expanding and new ones are being built, accounting for over 90 per cent of global urbanization (Datta and Shaban, 2016; Herbert and Murray, 2015; Wolff and Wiechmann, 2018; Zhang, 2016). In Africa, for example, the urban population is expected to triple by 2050, and the environmental costs of the extension of the built environment are already being paid with biodiversity loss and the depletion of regional ecosystems (Baldyga et al., 2008; Côté-Roy and Moser, 2019; Güneralp et al., 2017; Van Noorloos and Kloosterboer, 2018; Were et al., 2013).

Second, there is the issue of the metabolism that maintains the growth of cities and the genesis of new urban settlements. Like a living organism, a city needs a plethora of substances which are broken down and assimilated to yield energy, build infrastructure and eliminate waste. It has been estimated that cities consume approximately 75 per cent of natural resources, including fossil fuels, metal ores, non-metallic minerals and biomass (Pincetl, 2017; UNEP, 2014, 2016). Urban spaces absorb these flows of energy and materials which do not necessarily originate from them since, as noted in the past by Aristotle, a city needs to import what it does not produce enough of. What was, in Aristotle's time, a *need* has now become a *dependence*. While the global urban population keeps growing, the United Nations Environmental Programme stresses that local resource scarcity makes cities reliant on imports and on the development of 'complex infrastructure systems to transport essentials such as water, food and energy' (UNEP, 2017: 4). The criticality, and unsustainability, lies in the scarce supply and finite nature of much of the resources needed by the contemporary city.

Furthermore, the metabolism of cities not only consumes, but also produces. Across different scales, cities are responsible for approximately 70 per cent of global carbon emissions, and impact on the atmosphere and weather systems, thereby contributing to climate change (Bai et al., 2018; IPCC, 2015; Moran et al., 2018; Sudmant et al., 2018). Overall, considering the weight that they impose on global environmental changes, cities can be seen as contributors, if not the main actors of the so-called *Anthropocene*: a new and contested geological era in which humans are the dominant force behind the shaping of climate and the environment (see Lewis and Maslin, 2015; Steffen et al., 2007). As Pincetl (2017) notes, the majority of humans now live in cities and the Anthropocene, as an age shaped by humans, can consequently be understood as an age of cities or, in the words of West (2017), as the *Urbanocene*. From a more philosophical perspective, if the nature of the human being, as posited by Aristotle (2000), is intrinsically and inescapably urban, being *anthropos* implies being urban and, on this basis, living in the age of the anthropos implies living in an urban age.

Somehow paradoxically, the age of the anthropos, also the age of the urban, is an era in which the anthropos is not safe. In cities like Beijing, due to urban pollution, it is estimated that the life expectancy of citizens is being reduced by an average of 15 years, while in urban India every year thousands of people die prematurely because of poor air quality (Ghude et al., 2016; Guo et al., 2013; Lelieveld et al., 2015). Many cities have become deadly. Although caution is needed to avoid generalization, evidence suggests that the city is not supporting what Aristotle (2000, 2004) calls *eudaimonia*: human flourishing intended as the process through which humans realize their inner potential, thus reaching a state of satisfaction and happiness. Studies on the geography of happiness, for example, indicate that city dwellers tend to manifest a chronic lack of happiness, and that vast and hyper dense cities, due to a combination of long commutes, pollution, harmful noise, excessive artificial light and lack of therapeutic spaces like parks and bodies of water, impact negatively on wellbeing (Okulicz-Kozaryn, 2015; Okulicz-Kozaryn and Mazelis, 2018). Furthermore, in the majority of cases, when a city promotes the flourishing of its inhabitants and their happiness, the process is uneven. As shown particularly in the field of urban political ecology, the same social, political and physical dimensions of the urban that were highlighted by Aristotle, are producing and reproducing injustice (Harvey, 2009; Heynen et al., 2005; Kaika, 2005). Urbanization is creating spaces where large segments of the population are politically underrepresented, have little or no access to basic resources (such as energy and food), and are unevenly exposed to the burdens of the Anthropocene (Bouzarovski and Petrova, 2015; Hodson and Marvin, 2010; Sonnino, 2016). These grave problems indicate that the way cities are currently being planned, governed and experienced is largely unsustainable, and it is out of this realization that a global impetus for alternative urban models is emerging.

Experimental urbanism and the ideal city

In this context of global urban concerns and challenges, urban experimentation has become a popular way to address the unsustainability of cities, through the development of supposedly alternative models of urbanization (Bulkeley et al., 2014b; Bulkeley and Castán Broto, 2013; Bulkeley et al., 2019; Caprotti and Cowley, 2017; Evans et al., 2016; Karvonen and van Heur, 2014; Raven et al., 2019). The argument advanced by those in favour of urban experiments is that the current canons of city-making are flawed and, as such, they must be replaced. In itself, this claim is not new and has been heard many times throughout the ages. The city has always been a site of experimentation (Evans, 2011). Urban history is full of characters like Victor Frankenstein which have tried, across different spaces and times, to develop novel urban equations, claiming to possess the formula for the ideal city.

The Renaissance, for instance, with its cultural and philosophical emphasis on the human being as a creature capable of controlling destiny, by shaping the surrounding social and physical environment, presents several examples of projects for ideal cities (Kruft, 1989). These were not simply attempts to create perfect built environments, but rather experiments using the built environment to create an ideal society (Rosenau, 1983). The aim was not to create a geometrically perfect and aesthetically beautiful urban space. Architecture, urban design and planning were instruments serving political philosophy. In the context of the ideal-city phenomenon, the word *ideal* has a double meaning with only a tenuous connection to aesthetics. *Ideal* as an adjective referring to the best and most desirable city and, most importantly, *ideal* as a set of ideas of society and politics, upon which the genesis of the city is based. On these terms, Sforzinda, a project for a new city developed by Italian architect and philosopher Filarete as part of his *Trattato di architettura* (*Treatise on Architecture*), in the second half of the fifteenth century, is emblematic (Figure 1.1). Sforzinda manifests the effort to combine an ideal urban form with an ideal social form, in order to create what in the mind of its developer was the most desirable city.

The master plan for Sforzinda is characterized by a marked regular layout. The city has a radio-centric structure with a radial scheme for streets and canals, and an orthogonal scheme for squares and public buildings (Calabi, 2001). The centre of the city is designed as a vast public space around which Filarete positioned politically prominent buildings, such as the palace of the prince and the mint. However, Filarete's *Trattato* does not deal only with the geometry of the city. It also specifies the type of society that Sforzinda wants to cultivate and represent. First, the author aimed to create a homogenous society, in the attempt to avoid social fractures. Every area of the city is, to this end, connected through 16 radii and a wide circular street linking 16 minor squares to each other. This urban design was not meant to be purely functional from a mobility perspective, but to connect all citizens by opening up the city and promoting social encounters. In terms of housing, the master plan was tailored around concerns about poverty and affordability, which the author addressed by designing houses and buildings of different sizes and costs. There is an explicit aspiration for inclusive design, as Filarete sought to include mountain dwellers (then largely excluded from cities) within the walls of his ideal city. Second, the new ideal city was designed to refine the conscience of its citizens and reach moral and civic perfection. In this sense, emblematic in Filarete's plan for Sforzinda is the establishment of the House of Vice and Virtue: a ten-story building with a brothel on the ground floor, lecture rooms in the middle and an academy at the top, supposed to guide the ascent of man from vice to virtue.

After Sforzinda, the quest for the ideal city has been attempted by many architects, philosophers, urban planners and politicians which, like Filarete, rejected the

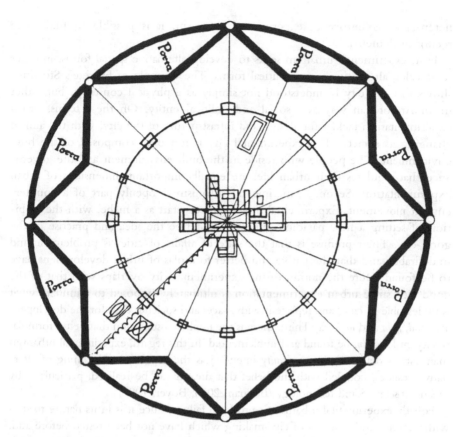

FIGURE 1.1 Master plan for Sforzinda (circa 1464).
Source: Wikimedia Commons

then mainstream models of city-making and proposed alternative urban equations. In the Enlightenment, a notable example is Chaux, a master-planned new town envisioned by French architect Claude-Nicolas Ledoux (1736–1806), to incarnate the avant-garde political ideas of Montesquieu and Rousseau (Kruft, 1989). Later emblematic attempts were made by Ebenezer Howard (1850–1928), Frank Lloyd Wright (1867–1959) and Le Corbusier (1887–1965), who brought new ideals and designs, giving different shapes to the ideal city (Fishman, 1982; Pinder, 2013). Howard's Garden City, for instance, sought to eliminate the barriers between the city and the countryside and blend their lifestyles within compact settlements, while in Wright's Broadacre the 'central belief was individualism' and houses were thus designed as independent and scattered units reachable by car and helicopter (Fishman, 1982: 94; Howard, 2007). The thread of urban experiments is long and cuts across many eras. Century after century, ideas have kept changing and so too the techniques, strategies and technologies employed to turn ideals into stone, and visions into actual cities. Yet, through the long and heterogeneous history of experimental urbanism, it is possible to find some recurring elements.

First, experimental urbanism seeks to develop alternative spatial forms, in order to develop alternative socio-political forms. The example of Filarete's Sforzinda shows that the city is understood not simply as a physical construct, but rather in an Aristotelian way as a social and political entity. On these terms, urban experimentation tackles the design and infrastructure of the city, with the aim of changing its society. The experimental city is not only composed of the built environment. The people who reside in the built environment and the governance that regulates their urban life, are equally important dimensions of urban experimentation. Second, experimental urbanism is openly part of a counter-current movement. Experimental urban projects start as a niche, with the ambition of setting a new paradigm meant to redefine the idea and practise of the good city. Their premise is that the state of affairs of cities is problematic and undesirable, and that alternative and better formulas of urban development have to be found. Here the notion of the experimental city overlaps with that of the good city, since urban experimentation is ultimately supposed to eliminate what is undesirable in cities and replace it with spaces and societies that, for the developers, are *good, right* and *desirable*. Third, there is the implicit assumption that better formulas of city-making *can* be found and implemented. In this regard, experimental urbanism manifests evident traits of *modernity* intended as the capacity of dreaming of alternative realities, coupled with the belief that dreams can be realized, particularly by means of science and technology (Berman, 2000; Boyer, 1997).

Fourth, experimental urbanism is prone to failure, since it is in its nature to deal with theories and practises of city-making which have not been tested before and, as such, are uncertain and potentially risky. However, an experimental urban project that fails can nonetheless be influential, eventually having a material impact on the built environment and an immaterial impact on the way urban planners, architects and policy-makers think about the city. The theory of the Garden City, for example, was never implemented exactly as it was originally envisioned by Ebenezer Howard, but it inspired and influenced the construction of numerous cities from around the world and its legacy is still standing (Hall, 2002). Moreover, even when an urban experiment collapses before reaching the implementation stage, future urban developers can learn from the failure of their predecessors, and integrate similar if not the same old ideas into new projects (Chang, 2017; Lovell, 2019; Temenos and Lauermann, 2020).

Fifth, urban experiments can be highly subjective visions. Sforzinda, Chaux, the Garden City and Broadacre, for instance, were proposed by singular individuals as the product of individual imaginations. Therefore, while experimental urbanism has the potential to advance alternative urban ideals, the outcome can be a new but narrow vision of the good city, since what is *good* and *bad, right* and *wrong, ideal* and *undesirable* is not defined by a collective intellectual inquiry or public political debate. This aspect is discernible not only from an imaginary and ethical perspective, in terms of what and whose ideas are taken into account and the value that is placed on them. The geographical and temporal perspectives might be narrow too in their looking exclusively at a specific space and time, thereby ignoring the applicability of the same ideal in different contexts, or simply assuming that a given ideal is universally and always valid and applicable. Urban experiments recurrently impose a rigid vision of the future, an *urban future* which clashes against the fluid and indefinite *yet to come*. The individualistic character of many experimental urban projects has also been, historically, one of the main reasons why large-scale urban experiments have rarely succeeded. In this sense, although distant in time and space, Sforzinda and Broadacre are related by the fact that their inventors failed to create or join a network capable of financing the projects. As a result, none of them was built and both fell into the realm of utopia (Fishman, 1982; Kruft, 1989).

Sixth, practitioners of experimental urbanism tend to test and implement their theories under controlled conditions (Evans and Karvonen, 2011). This typically means that the experiment starts with a city, or part of it, whose development is monitored and regulated. The objective is usually to scale up the experiment at a later stage, if it is successful. This is an aspect of experimental urbanism which indicates the diverse scales of urban experimentation. Small-scale urban experiments can target a district or even just a single building, and then potentially extend their influence to the whole city. A large-scale urban experiment might involve building a new city from scratch, which can subsequently serve as a model for the construction of similar cities across the country. Along this spectrum ranging from modest interventions to grandiose mega-projects, while the scale is different, the ethos of experimentation and the will to change the built environment in order to trigger broader social and political changes remain. Seventh, the implementation of experimental urban projects is frequently disciplined by a master plan which is supposed to provide developers with a scientific methodology (Cugurullo, 2018). Master plans set and arrange the steps necessary to complete urban experiments, determining what has to be built, how, where and when. This is often the ground upon which advocates of experimental urbanism, claim that this typology of city-making differs from a standard and more chaotic process of urban development.

Finally, and arguably most importantly, all the points above can be *false*. There can be strong differences between what the developers and promoters of a supposedly experimental urban project claim and what is actually happening on the ground. In classical and modern philosophy, this is commonly understood as *correspondence theory*, the idea that a statement does not necessarily correspond to a fact (see Kirkham, 1992). History abounds with allegedly experimental, innovative and ideal cities whose reality was far from what developers and stakeholders had claimed. Valletta (Malta), for instance, whose construction (1566–1573) was co-financed by Pope Pius IV and various members of the European aristocracy, in theory to incarnate the ideal of Christianity into a city, was in reality a fortress meant to keep Ottoman pirates at bay, and preserve trade in the Mediterranean Sea (Kruft, 1989). There was thus a stark discrepancy between the discourses through which Valletta was being promoted (a city supposed to protect the ideas and values of Christianism) and the actual Valletta (a city protecting long-standing power relations and politico-economies). Here the adjective *false* describes an urban project that is promoted as experimental, counter-current and as a medium to realize certain ideals, while the facts show otherwise. In the twenty-first century, critical scholars working in the field of experimental urbanism have been investigating the extent to which contemporary experimental urban projects are actually driving real change and achieving urban sustainability, questioning the assumptions of what are promoted as, but not necessarily are, *experimental cities* (Castán Broto and Bulkeley, 2013; Cugurullo, 2016; Kaika, 2017; Karvonen et al., 2014; McGuirk et al., 2014; Savini and Bertolini, 2019).

This book focuses on the most popular and influential typologies of experimental urbanism of the twenty-first century: the *eco-city*, the *smart city* and the emerging *autonomous city* run not by *human* but by *artificial* intelligences. It examines the theories behind their genesis and assesses their implementation, evaluating the extent to which these supposedly experimental and sustainable urban projects are achieving sustainability. The story of eco, smart and autonomous cities is not linear, and here lies a key difference between this text and its literary guiding spirit, *Frankenstein*. While profound, intellectually sophisticated and enriched by the perspectives of different characters, Mary Shelley's book follows a fairly linear narrative. *Frankenstein* has a protagonist, several secondary characters and an antagonist. The protagonist embarks on a quest which starts from a clear and familiar context, to then push the story into the unknown. This book has a plethora of protagonists. Many of them are hidden, and do not even have a human face. The context in which they operate is, from the very beginning, ambiguous and tends to become amorphous step by step. The following chapters are an attempt to give a narrative to the dubious development of eco-cities, the hazy creation of smart cities and their complex evolution into cities controlled by enigmatic artificial intelligences.

Methodology

Giving a narrative is about ordering seemingly disparate and obscure events into a coherent story. The narrative of this book unfolds through a case-study approach. From an empirical point of view, most of the focus is on the analysis of two cities: Masdar City in Abu Dhabi as an example of a new eco-city project, and Hong Kong as an instance of a large-scale smart-city initiative. The rationale behind the choice of Masdar City and Hong Kong is twofold. This study does not aim to offer a comparative analysis, but rather a detailed and empirically rich examination of contemporary urban experiments, first in the two mainstream typologies of experimental urbanism (eco-cities and smart cities) and second in the two main types of built environment (new cities and existing settlements). Fieldwork was conducted, at different stages, in Abu Dhabi and Hong Kong across 2010 and 2016, for a total of 18 months. Regarding the autonomous city, the phenomenon of built environments operated and governed by *urban artificial intelligences* is an emerging one and, to date, there is scarce empirical ground for in-depth case-study research (Cugurullo, 2020). Autonomous cities are crossing the frontiers of urban experimentation, entering multiform territories which are largely uncharted. Therefore, the book explores the autonomous city by drawing upon diverse real-life examples, with the aim of unveiling the heterogeneous and complex spectrum of the use of artificial intelligence (AI) in cities, rather than analyzing one individual aspect in detail.

Much of the information disclosed during the research on Masdar City and Hong Kong's smart-city agenda is controversial in nature and not publicly available. A total of 35 semi-structured and 23 unstructured interviews were conducted with members of the public sector, such as policy-makers, developers and spatial planners from local planning councils, as well as representatives from architecture firms, investment companies and clean-tech multinationals. In addition, key documents, including master plans, development agendas and environmental reports, were examined to triangulate the information that emerged in the interviews. The data provides evidence of the many problems which undermine the sustainability of the two projects, clashing with the claims of developers and stakeholders. In order to protect the anonymity of the participants, across the empirical chapters all the names of the interviewees have been replaced by their role and position. Following the same ethical considerations, the book does not refer directly to the documents that are not public, as this would expose the identity of those who shared them.

From a theoretical point of view, the book employs its empirical basis to propose general theories and critiques of experimental urbanism. As Flyvbjerg (2006) remarks, in-depth case studies can provide precious insights into broader phenomena, thus turning specific information into general knowledge. Along this line of thought, the specific cases of Masdar City and Hong Kong serve the purpose of capturing general trends in urban experiments such as, for example, the emergence of AI in the governance of cities. However, given the limitations that are intrinsic to case-study research, the objective of the book is not to provide rigid one-size-fits-all conceptual frameworks. On these terms, the following arguments are animated by a philosophy of research akin to what Peck (2017a, 2017b) defines as *conjunctural urbanism*. The book approaches contemporary experimental urban projects, first as part of a broader and much older trend in urban development since, across history, cities have been recurrently used as vehicles to experiment with alternative forms of social organization. Second, urban experimentation is here understood as a situated and diverse phenomenon connected to the specificity of the single case studies. Therefore, by approaching the subject of inquiry as the interconnection or conjuncture of these two dimensions, the book seeks to offer a 'midlevel formulation' whose explanatory power and generalizability remain open and revisable (Peck, 2017a: 19–20).

Structure of the book

The book is divided into three parts which mirror the unfolding of the events narrated in Mary Shelley's novel. In the first part, *The literature*, the focus is on the key ideas and theories of ecological urbanism and smart urbanism. Before delving into his experiment, Victor spends several years studying, reading books and absorbing the literature. He knows that what he is about to empirically attempt has already been the subject of many studies. Conscious that there is a lot to learn from the scholars that came before him, Victor reviews and tries to make sense of a very heterogeneous literature. Given the complexity of his field of research, the young scientist engages with a broad spectrum of disciplines ranging from chemistry to physiology. In so doing, he learns about the principle of life, how a human being becomes such, and how the human body can, in theory, be perfected. Following this narrative, in Chapters 2 and 3, the book reviews and discusses the literature on eco-cities and smart cities, as a way to understand the conceptual foundations of eco-city and smart-city projects. Like in the case of Victor Frankenstein, this is not just a literature review. There is not a single source or discipline behind the notions of the eco-city and the smart city. Instead, there are fragmented ideas, visions and images coming from diverse and, at times, contrasting fields of knowledge. The first part of the book, therefore, seeks to connect the dots or, put differently, the threads that, when woven together, form the principles of eco-city-making (Chapter 2) and smart-city-making (Chapter 3). This is a journey cutting across different branches of environmental philosophy, planning theory, urban design, political science, political philosophy, geography and more. By the end of Part I, the book aims to have clarified the often obscure and misused terms *eco-city* and *smart city*, to then move to their empirical incarnations.

In the second part of *Frankenstein*, Mary Shelley describes the empirical work of Victor Frankenstein. After having spent a considerable time surrounded by books, the Doctor surrounds himself with a complex array of tools, machines and materials. He is ready (or so he believes) to finally conduct his experiment and put the theory into practise. Victor locks himself up in a laboratory and the experiment begins. In the second part of this book, *The experiment*, the focus is on the practise of ecological urbanism and smart urbanism. Chapter 4 explores an actually existing project for a new eco-city, while Chapter 5 investigates the implementation of a smart-city agenda. These two chapters form the empirical core of the book. Data collected in the field is here used to examine where, how, why, for whom and by whom experimental projects for eco and smart cities are developed. The book seeks to shed light not simply on what is happening on the ground, but also on the discrepancy between the ideas of eco-city and smart city and their empirical incarnations. The empirics cover different scales and manifestations of the urban as it is shaped by eco and smart-city initiatives. The lens of inquiry moves from single buildings to districts, and from the entire city to the surrounding region, thereby revealing the many facets of being *eco* and *smart* in an urban context.

The third part of Mary Shelley's novel deals with the consequences of the experiment conducted by Victor. Eventually, the Doctor has to face what he has created and confronts the repercussions of his actions. The result of the experiment is a creature which is alive and independent. It has its own agency. It acts. It evolves, turning into something which Victor had not expected before he started experimenting. Ultimately, for him, the outcome of the experiment comes as a tragic revelation or, in other words, an apocalypse. In the final part of this book, *The apocalypse*, the narrative unveils the results of the urban experiments developed through eco-city and smart-city initiatives. Chapter 6 problematizes the sustainability challenges produced by alleged eco-cities and smart cities, and emphasizes their fragmented and dysfunctional character, by using Frankenstein's monster as a metaphor. Chapter 7 investigates the evolution of experimental urban projects in light of recent advancements in AI which are introducing autonomous technologies in the management of cities. Here the book depicts an emerging autonomous city: a space, born out of years of eco and smart-city experiments, where diverse artificial intelligences, from service robots to digital platforms, perform urban activities that have traditionally been human activities. Chapter 8 consists of an epilogue looking at the possible urban futures that lie ahead, too far for being now real, but not too far for being realizable. The scenario that is presented shows cities radically altered by AI to the point of losing those key qualities and characteristics that make them cities in the first place. The end of the book is about the end of the city, intended not as a global calamity causing the destruction of urban spaces, but rather as the termination of the city as a place predominately governed, planned and experienced by human intelligences.

While this book deals with the future, it does not try to predict it and its analysis of past and present urban experiments is meant to understand and evaluate the directions that urban development is currently taking. Directions that are not carved in stone and that should and can be changed. The tale of Frankenstein is a warning. Frankenstein's experiment gets out of control and its results are deadly. Blinded by hubris, Victor irresponsibly creates and then abandons a powerful being whose integration into human society proves to be disastrous. Similarly, the tale that follows shows how in the passage from eco and smart cities to the autonomous city, a reckless technological development is producing urban spaces which humans barely control, scarcely understand and might not be compatible with. This tale is a warning too. Cities are heading toward a dangerous future and must be careful. Sooner or later, humanity will have to respond to the outcomes of experimental urbanism. Victor's response to the monstrosity of technological experimentation is *hatred*. He gives up science and finds in a gun the answer to his problems, in the attempt to kill the monster that he has created. This book's response is *love*. Love as making an effort to understand the numerous human and non-human intelligences that populate cities, by using the tools of the social sciences and humanities. Artificial intelligences, in particular, since their arcane explainability is a barrier preventing many people from understanding and trusting them (Barredo Arrieta et al., 2020; Stoyanovich et al., 2020). Love as politics here is intended as an invitation to actively engage with urban experiments as a community of informed citizens, sharing ideas, debating, voting, protesting when necessary and striving to include missing ideals of justice and ecology in the engine of experimental urbanism. Victor fails his experiment because he fails to know, to empathize and to engage. The following pages provide the knowledge that Victor was lacking in the hope of stimulating empathy for and engagement with the urban experiments that are now leaving the laboratory and entering the everyday.

References

Angelo, H. and Wachsmuth, D. (2020). Why does everyone think cities can save the planet? *Urban Studies*, 57 (11), 2201–2221.

Aristotle (1992). *The Politics*. Penguin Classics, London.

Aristotle (1996). *Physics*. Penguin Classics, London.

Aristotle (2000). *The Politics*. Oxford University Press, Oxford.

Aristotle (2004). *The Nicomachean ethics*. Penguin Classics, London.

Aristotle (2005). *La politica*. Laterza, Roma-Bari.

Aristotle (2013). *The Politics*. The University of Chicago Press, Chicago.

Barredo Arrieta, A., Díaz-Rodríguez, N., Del Ser, J., Bennetot, A., Tabik, S., Barbado, A., Garcia, S., Gil-Lopez, S., Molina, D., Benjamins, R., Chatila, R. and Herrera, F. (2020). Explainable Artificial Intelligence (XAI): Concepts, taxonomies, opportunities and challenges toward responsible AI. *Information Fusion*, 58, 82–115.

Bai, X., Dawson, R. J., Ürge-Vorsatz, D., Delgado, G. C., Barau, A. S., Dhakal, S., Dodman, D., Leonardsen, L., Masson-Delmotte, V., Roberts, D. and Schultz, S. (2018). Six research priorities for cities and climate change. *Nature*, 555 (7694), 23–25.

Baldyga, T. J., Miller, S. N., Driese, K. L. and Gichaba, C. M. (2008). Assessing land cover change in Kenya's Mau Forest region using remotely sensed data. *African Journal of Ecology*, 46 (1), 46–54.

Beloin-Saint-Pierre, D., Rugani, B., Lasvaux, S., Mailhac, A., Popovici, E., Sibiude, G., Benedetto, E. and Schiopu, N. (2017). A review of urban metabolism studies to identify key methodological choices for future harmonization and implementation. *Journal of Cleaner Production*, 163, S223–S240.

Bettencourt, L. and West, G. (2010). A unified theory of urban living. *Nature*, 467 (7318), 912.

Blundell, S. (1995). *Women in ancient Greece*. Harvard University Press, Cambridge.

Bouzarovski, S. and Petrova, S. (2015). A global perspective on domestic energy deprivation: Overcoming the energy poverty–fuel poverty binary. *Energy Research & Social Science*, 10, 31–40.

Boyer, M. C. (1997). *Dreaming the rational city: The myth of American city planning*. MIT Press, Cambridge.

Brenner, N. *(Ed.)*. (2014). *Implosions/explosions: Towards a study of planetary urbanization*. Jovis Verlag, Berlin.

Brenner, N. (2019). *New urban spaces: Urban theory and the scale question*. Oxford University Press, Oxford.

Bulkeley, H. and Castán Broto, V. (2013). Government by experiment? Global cities and the governing of climate change. *Transactions of the Institute of British Geographers*, 38 (3), 361–375.

Bulkeley, H. A., Castán Broto, V. and Edwards, G. A. (2014a). *An urban politics of climate change: experimentation and the governing of socio-technical transitions*. Routledge, London.

Bulkeley, H., Castán Broto, V. and Maassen, A. (2014b). Low-carbon transitions and the reconfiguration of urban infrastructure. *Urban Studies*, 51 (7), 1471–1486.

Bulkeley, H., Marvin, S., Palgan, Y. V., McCormick, K., Breitfuss-Loidl, M., Mai, L., von Wirth, T. and Frantzeskaki, N. (2019). Urban living laboratories: Conducting the experimental city?. *European urban and regional studies*, 26 (4), 317–335.

Berman, M. (2000). *All that is solid melts into air: the experience of modernity*. Verso, London.

Calabi, D. (2001). *La citta' del primo Rinascimento*. Laterza, Roma-Bari.

Caprotti, F. and Cowley, R. (2017). Interrogating urban experiments. *Urban Geography*, 38 (9), 1441–1450.

Carr Rider, B. (2014). *The Greek house; its history and development from the Neolithic period to the Hellenistic age*. Cambridge University Press, Cambridge.

Castán Broto, V. and Bulkeley, H. (2013). A survey of urban climate change experiments in 100 cities. *Global Environmental Change*, 23 (1), 92–102.

Chang, I. C. C. (2017). Failure matters: Reassembling eco-urbanism in a globalizing China. *Environment and Planning A*, 49 (8), 1719–1742.

Chen, G., Li, X., Liu, X., Chen, Y., Liang, X., Leng, J., Xu, X., Liao, W., Qiu, Y., Wu, Q. and Huang, K. (2020). Global projections of future urban land expansion under shared socioeconomic pathways. *Nature communications*, 11 (1), 1–12.

Conforti, C. (2005). *La citta' del tardo Rinascimento*. Laterza, Roma.

Conke, L. S. and Ferreira, T. L. (2015). Urban metabolism: Measuring the city's contribution to sustainable development. *Environmental pollution*, 202, 146–152.

Côté-Roy, L. and Moser, S. (2019). 'Does Africa not deserve shiny new cities?' The power of seductive rhetoric around new cities in Africa. *Urban Studies*, 56 (12), 2391–2407.

Cugurullo, F. (2016). Urban eco-modernisation and the policy context of new eco-city projects: Where Masdar City fails and why. *Urban Studies*, 53 (11), 2417–2433.

Cugurullo, F. (2018). Exposing smart cities and eco-cities: Frankenstein urbanism and the sustainability challenges of the experimental city. *Environment and Planning A: Economy and Space*, 50 (1), 73–92.

Cugurullo, F. (2020). Urban artificial intelligence: from automation to autonomy in the smart city. *Frontiers in Sustainable Cities*, doi:10.3389/frsc.2020.00038.

Datta, A. and Shaban, A. (Eds.). (2016). *Mega-urbanization in the global south: Fast cities and new urban utopias of the postcolonial state*. Routledge, London.

Dreizehnter, A. (1970). *Aristotles's Politik*. Fink, Munich.

Elmqvist, T., Bai, X., Frantzeskaki, N., Griffith, C., Maddox, D., McPhearson, T., Parnell, S., Romero-Lankao, P., Simon, D. and Watkins, M. (Eds.). (2018). *The urban planet: Knowledge towards sustainable cities*. Cambridge University Press, Cambridge.

Evans, J. P. (2011). Resilience, ecology and adaptation in the experimental city. *Transactions of the Institute of British Geographers*, 36 (2), 223–237.

Evans, J. and Karvonen, A. (2011). Living laboratories for sustainability: exploring the politics and epistemology of urban transition. In Bulkeley, H., Castán Broto, V., Hodson, M. and Marvin, S. (Eds.). *Cities and low carbon transitions*. Routledge, London, pp. 126–141.

Evans, J., Karvonen, A. and Raven, R. (Eds.). (2016). *The experimental city*. Routledge, London.

Fishman, R. (1982). *Urban utopias in the twentieth century: Ebenezer Howard, Frank Lloyd Wright, and Le Corbusier*. MIT Press, Cambridge.

Flyvbjerg, B. (2006). Five misunderstandings about case-study research. *Qualitative inquiry*, 12 (2), 219–245.

Ghude, S. D., Chate, D. M., Jena, C., Beig, G., Kumar, R., Barth, M. C., Fadnavis, S. and Pithani, P. (2016). Premature mortality in India due to PM2. 5 and ozone exposure. *Geophysical Research Letters*, 43 (9), 4650–4658.

Güneralp, B., Lwasa, S., Masundire, H., Parnell, S. and Seto, K. C. (2017). Urbanization in Africa: challenges and opportunities for conservation. *Environmental Research Letters*, 13 (1), 015002.

Guo, Y., Li, S., Tian, Z., Pan, X., Zhang, J. and Williams, G. (2013). The burden of air pollution on years of life lost in Beijing, China, 2004–08: retrospective regression analysis of daily deaths. *BMJ*, 347, f7139.

Hall, P. (2002). *Cities of tomorrow*. Blackwell Publishers, Oxford

Hansen, T. and Coenen, L. (2015). The geography of sustainability transitions: Review, synthesis and reflections on an emergent research field. *Environmental innovation and societal transitions*, 17, 92–109.

Harding, A. and Blokland, T. (2014). *Urban theory: a critical introduction to power, cities and urbanism in the 21st century*. Sage, London.

Harvey, D. (2009). *Social justice and the city*. University of Georgia Press, Athens and London.

He, C., Liu, Z., Tian, J. and Ma, Q. (2014). Urban expansion dynamics and natural habitat loss in China: a multiscale landscape perspective. *Global change biology*, 20 (9): 2886–2902.

Herbert, C. W. and Murray, M. J. (2015). Building from scratch: new cities, privatized urbanism and the spatial restructuring of Johannesburg after apartheid. *International Journal of Urban and Regional Research*, 39 (3), 471–494.

Heynen, N., Kaika, M. and Swyngedouw, E. (Eds.) (2005). *In the nature of cities. Urban political ecology and the politics of urban metabolism*. Routledge, London.

Hodson, M. and Marvin, S. (2010). Urbanism in the Anthropocene: Ecological urbanism or premium ecological enclaves? *City*, 14 (3), 298–313.

Howard, E. (2007). *Garden cities of to-morrow*. Routledge, London.

Iossifova, D., Doll, C. N. and Gasparatos, A. (Eds.). (2018). *Defining the Urban: Interdisciplinary and Professional Perspectives*. Routledge, London.

IPCC (2015). *Climate change 2014: Mitigation of climate change*. Cambridge University Press, New York.

Jayne, M. and Ward, K. (Eds.). (2016). *Urban theory: New critical perspectives*. Routledge, London.

Kaika, M. (2005). *City of flows: Modernity, nature, and the city*. Routledge, London.

Kaika, M. (2017). 'Don't call me resilient again!': the New Urban Agenda as immunology… or… what happens when communities refuse to be vaccinated with 'smart cities' and indicators. *Environment and Urbanization*, 29 (1), 89–102.

Karvonen, A. and van Heur, B. (2014). Urban laboratories: experiments in reworking cities. *International Journal of Urban and Regional Research*, 38, 379–392.

Karvonen, A., Evans, J. and van Heur, B. (2014). *The politics of urban experiments: radical change or business as usual?* In Marvin, S. and Hodson, M. (Eds.). *After Sustainable Cities*. Routledge, London, pp. 105–114.

Kirkham, R. L. (1992). *Theories of truth: A critical introduction*. MIT Press, Cambridge.

Kruft, H. W. (1989). *Städte in Utopia: die Idealstadt vom 15. bis zum 18. Jahrhundert zwischen Staatsutopie und Wirklichkeit*. CH Beck, Munich.

Lefebvre, H. (2009). *The production of space*. Blackwell, Oxford.

Lefebvre, H. (2014/1987). *Dissolving city, planetary metamorphosis*. In Brenner, N. (Ed.). *Implosions/Explosions: Towards a Study of Planetary Urbanization*. Jovis Verlag, Berlin, pp. 566–570.

Lelieveld, J., Evans, J. S., Fnais, M., Giannadaki, D. and Pozzer, A. (2015). The contribution of outdoor air pollution sources to premature mortality on a global scale. *Nature*, 525 (7569), 367.

Lewis, S. L. and Maslin, M. A. (2015). Defining the Anthropocene. *Nature*, 519 (7542), 171.

Li, H., Wei, Y. D., Liao, F. H. and Huang, Z. (2015). Administrative hierarchy and urban land expansion in transitional China. *Applied Geography*, 56, 177–186.

Lovell, H. (2019). Policy failure mobilities. *Progress in Human Geography*, 43 (1), 46–63.

McGuirk, P., Dowling, R. and Bulkeley, H. (2014). Repositioning urban governments? Energy efficiency and Australia's changing climate and energy governance regimes. *Urban Studies*, 51 (13), 2717–2734.

McNeill, D. (2016). *Global cities and urban theory*. Sage, London.

Melchiorri, M., Florczyk, A. J., Freire, S., Schiavina, M., Pesaresi, M. and Kemper, T. (2018). Unveiling 25 Years of Planetary Urbanization with Remote Sensing: Perspectives from the Global Human Settlement Layer. *Remote Sensing*, 10 (5), 768.

Morachiello, P. (2004). *La città greca*. Laterza, Roma-Bari.

Moran, D., Kanemoto, K., Jiborn, M., Wood, R., Többen, J. and Seto, K. C. (2018). Carbon footprints of 13 000 cities. *Environmental Research Letters*, 13 (6), 064041.

Mumford, L. (1961). *The city in history: Its origins, its transformations, and its prospects*. Harcourt, Brace & World, New York.

Okulicz-Kozaryn, A. (2015). *Happiness and place: Why life is better outside of the city*. Springer, New York.

Okulicz-Kozaryn, A. and Mazelis, J. M. (2018). Urbanism and happiness: A test of Wirth's theory of urban life. *Urban Studies*, 55 (2), 349–364.

Peake, L., Patrick, D., Reddy, R. N., Sarp Tanyildiz, G., Ruddick, S. and Tchoukaleyska, R. (2018). Placing planetary urbanization in other fields of vision. *Environment and Planning D: Society and Space*, 36 (3), 374–386.

Peck, J. (2017a). Transatlantic city, part 1: Conjunctural urbanism. *Urban Studies*, 54 (1), 4–30.

Peck, J. (2017b). Transatlantic city, part 2: Late entrepreneurialism. *Urban Studies*, 54 (2), 327–363.

Pincetl, S. (2017). Cities in the age of the Anthropocene: Climate change agents and the potential for mitigation. *Anthropocene*, 20, 74–82.

Pincetl, S., Bunje, P. and Holmes, T. (2012). An expanded urban metabolism method: Toward a systems approach for assessing urban energy processes and causes. *Landscape and urban planning*, 107 (3), 193–202.

Pinder, D. (2013). *Visions of the city: Utopianism, power and politics in twentieth century urbanism*. Routledge, London.

Raven, R., Sengers, F., Spaeth, P., Xie, L., Cheshmehzangi, A. and de Jong, M. (2019). Urban experimentation and institutional arrangements. *European Planning Studies*, 27 (2), 258–281.

Rode, P., Terrefe, B. and da Cruz, N. F. (2020). Cities and the governance of transport interfaces: Ethiopia's new rail systems. *Transport Policy*, 91, 76–94.

Rosenau, H. (1983). *The Ideal City in its Architectural Evolution*. Methuen & Co, London.

Ross, W. D. (1957). *Aristotelis politica*. Oxford Classical Texts, Oxford.

Savini, F. and Bertolini, L. (2019). Urban experimentation as a politics of niches. *Environment and Planning A: Economy and Space*, 0308518X19826085.

Schmid, C. (2018). Journeys through planetary urbanization: Decentering perspectives on the urban. *Environment and Planning D: Society and Space*, 36 (3), 591–610.

Scruton, R. (2017). *On human nature*. Princeton University Press, Princeton.

Sonnino, R. (2016). The new geography of food security: exploring the potential of urban food strategies. *The Geographical Journal*, 182 (2), 190–200.

Steffen, W., Crutzen, P. J. and McNeill, J. R. (2007). The Anthropocene: are humans now overwhelming the great forces of nature. *AMBIO: A Journal of the Human Environment*, 36 (8), 614–621.

Stoyanovich, J., Van Bavel, J. J. and West, T. V. (2020). The imperative of interpretable machines. *Nature Machine Intelligence*, 2 (4), 197–199.

Sudmant, A., Gouldson, A., Millward-Hopkins, J., Scott, K. and Barrett, J. (2018). Producer cities and consumer cities: Using production-and consumption-based carbon accounts to guide climate action in China, the UK, and the US. *Journal of Cleaner Production*, 176, 654–662.

Temenos, C. and Lauermann, J. (2020). The urban politics of policy failure. *Urban Geography*, 1–10.

Truffer, B., Murphy, J. T. and Raven, R. (2015). The geography of sustainability transitions: Contours of an emerging theme. *Environmental Innovation and Societal Transitions*, 17, 63–72.

UNEP (2014). Annual report 2013. [Online] Available: http://wedocs.unep.org/bitstream/handle/20.500.11822/8607/-UNEP%202013%20Annual%20Report-2014UNEP%20AR%202013-LR.pdf?sequence=8&isAllowed=y [Accessed 10 November 2020].

UNEP (2016). Global material flows and resource productivity. [Online] Available: http://wedocs.unep.org/bitstream/handle/20.500.11822/21557/global_material_flows_full_report_english.pdf?sequence=1&isAllowed=y [Accessed 10 November 2020].

UNEP (2017). Resilience and resource efficiency in cities. [Online] Available: https://wedocs.unep.org/bitstream/handle/20.500.11822/20629/Resilience_resource_efficiency_cities.pdf?sequence=1&isAllowed=y [Accessed 10 November 2020].

Van Noorloos, F. and Kloosterboer, M. (2018). Africa's new cities: The contested future of urbanisation. *Urban Studies*, 55 (6), 1223–1241.

Vitz, M. (2018). *A city on a lake: Urban political ecology and the growth of Mexico City*. Duke University Press, Durham.

Wang, T., Tian, X., Hashimoto, S. and Tanikawa, H. (2015). Concrete transformation of buildings in China and implications for the steel cycle. *Resources, Conservation and Recycling*, 103, 205–215.

Were, K. O., Dick, Ø. B. and Singh, B. R. (2013). Remotely sensing the spatial and temporal land cover changes in Eastern Mau forest reserve and Lake Nakuru drainage basin, Kenya. *Applied Geography*, 41, 75–86.

West, G. (2017). *Scale: The universal laws of growth, innovation, sustainability, and the pace of life in organisms, cities, economies, and companies*. Penguin Press, London

Whitehead, M. (2003). (Re) analysing the sustainable city: Nature, urbanisation and the regulation of socio-environmental relations in the UK. *Urban Studies*, 40 (7), 1183–1206.

Whitehead, M. (2007). *Spaces of sustainability: geographical perspectives on the sustainable society*. Routledge, London.

Williams, J., Robinson, C. and Bouzarovski, S. (2020). China's Belt and Road Initiative and the emerging geographies of global urbanisation. *The Geographical Journal*, 186 (1), 128–140.

Wilson, D. and Jonas, A. E. (2018). Planetary urbanization: new perspectives on the debate. *Urban Geography*, 39 (10), 1576–1580.

Wolff, M. and Wiechmann, T. (2018). Urban growth and decline: Europe's shrinking cities in a comparative perspective 1990–2010. *European Urban and Regional Studies*, 25 (2), 122–139.

Zhang, X. Q. (2016). The trends, promises and challenges of urbanisation in the world. *Habitat International*, 54, 241–252.

Zhang, T., Chen, S. S. and Li, G. (2020). Exploring the relationships between urban form metrics and the vegetation biomass loss under urban expansion in China. *Environment and Planning B: Urban Analytics and City Science*, 47 (3), 363–380.

PART I
The literature

'But here were books, and here were men who had penetrated deeper and knew more.'

(Frankenstein, *Chapter 2*)

2

THEORIES OF ECOLOGICAL URBANISM

Introduction

Although eco-city initiatives have become popular in the twenty-first century, visions of cities in balance with the natural environment began to circulate many centuries before then. What these visions represent is far from being homogeneous. The *eco-city* concept escapes a universal definition, and its theoretical underpinnings are hard to grasp and weave into a coherent model of city-making. The aim of this chapter is to explore and discuss the key ideas which form the theory of *ecological urbanism* in search of the conceptual origins of the term *eco-city*, whose material incarnations will be discussed in the second part of the book. This is a journey which will cut across time and space, showing where ideas of ecological urbanism come from, in order to understand what urban futures they might lead to. Rather than providing a dogmatic definition of what an eco-city is and a *modus operandi* for its realization, the following sections seek to pull together, from different sets of literature, the recurring themes in eco-city studies, discourses, contemplations and, at times, speculations. The ideas, visions and images collected will then be assembled into a mosaic which, although by nature fragmented, is an attempt to show what is commonly meant when the term *eco-city* is evoked.

The above aim necessitates a short premise. The theoretical landscape of the eco-city is extremely broad. Its boundaries are porous and not clearly defined and, as a result, ecological urbanism often trespasses on the realm of other urbanisms, such as *sustainable* urbanism, *smart* urbanism and *resilient* urbanism. This, in turn, creates a lot of theoretical confusion, since the juxtaposition of the terms *ecological, sustainable, smart* and *resilient* conceals what is special about ecological urbanism and what *ecological* refers to in ecological urbanism (Hagan, 2014; Mostafavi and Doherty, 2016). Adding to this theoretical confusion is a twofold factual problem. First, the term *ecological* is per se ambiguous and so is its application in urban development

(Gandy, 2015). Second, those existing urban projects that are in theory supposed to realize an ecological urbanism, so-called *eco-cities*, have been widely critiqued for causing numerous socio-environmental problems and misinterpreting the notion of ecological urbanism (Caprotti, 2014; Chang and Sheppard, 2013; Cugurullo, 2016). It is therefore crucial to clarify the connotation of ecological urbanism because this urbanism, so will this chapter contend, has its own distinct meaning as well as a powerful message which is of value to contemporary cities.

The chapter will examine both the potential benefits of ecological urbanism and its limitations. The eco-city, it is important to clarify, is far from being a perfect ideal. It is not the product of a complete and coherent theoretical system and, like all ideals, the quality of being perfect is highly subjective and context-dependent. As the following sections illustrate, ecological urbanism has been conceptualized through the ages in different ways according to different aspects of urban development, such as the physical structure of cities, their design and governance. When taken individually, single dimensions of ecological urbanism can manifest severe intellectual gaps which, in turn, might render the whole *eco-city* concept barely credible. Building upon this premise, in addition to tracing the ideological origins of the eco-city, this chapter also contributes to the development of a holistic body of principles of ecological urbanism. It draws upon complementary lessons from urban geography, urban planning, urban design, urban sociology, urban history, political philosophy and political science, to capture the multiple dimensions of the eco-city. Some of the theoretical insights that are here taken into account, are not directly part of the landscape of knowledge upon which eco-cities usually sit. However, as this chapter will argue, ecological urbanism and, therefore, the eco-city are not just frameworks meant to discipline urbanization with the laws of ecology. They are also categories of social and political thought and, as such, must engage with literature going well beyond the natural sciences.

The remainder of the chapter is structured as follows. First, the narrative delves into ecological urbanism with a focus on planning theory. The chapter introduces the concept of *eco-cities* by drawing upon the work of Richard Register whose thinking establishes a direct thematic connection between urban planning and urban ecology. The planning of eco-cities is approached through the science of ecology which, in ecological urbanism, emerges as a medium to build cities in balance with local and regional ecosystems. As the chapter shows, in ecological urbanism the boundaries of urbanization are found in the geographical and geological landmarks that serve as the natural perimeter of a city. The limitations of this theoretical approach to urban planning are tackled via Léon Krier's studies in which urbanization is imagined as *duplication*. Rather than as a capitalist impetus of perpetual growth, this strand of ecological urbanism sees urban development as a finite process which stops in a given place while starting anew somewhere else: a theory resonating with biological research on budding. Second, the chapter covers the design of eco-cities using a twofold understanding of ecology: the science of ecology as a design tool utilized to reduce the amount of built environment reserved for cars in order to maximize the amount of green space in cities; and also ecology as a metaphor for social diversity. In this section, the book remarks that ecological urbanism is about preserving the biodiversity of a geographical area, as well as cultivating the diversity of the people who populate it. In this sense, the *eco-city* concept evokes images of cities where the heterogeneous needs and desires of citizens are translated into diverse, accessible and compact urban spaces. Finally, the chapter traverses the domain of political philosophy, looking at theories of governance with an explicit spatial emphasis on socio-environmental issues. Using Spinoza as a starting point, the book concentrates the discussion on social ecology and eco-socialism to stress how the understanding and, therefore, the realization of eco-cities is a quest which demands more than the redesign of the physical structure of cities. An ecological urbanism will not be realized, in theory and practise, unless its political context is intellectually challenged.

Planning the eco-city

From a nominal point of view, the story begins with Richard Register, an American environmental activist with a background in architecture, who coined the term *eco-city* in 1987. This is the year of Register's seminal publication *Ecocity Berkeley*: a short book which gave international momentum to ecological urbanism. Prior to writing *Ecocity Berkeley*, during the 1970s Register was acting locally in California as part of a non-profit organization called *Urban Ecology*, in the attempt to improve the environmental conditions of Berkeley, by practising agriculture on the streets and promoting cycling (Roseland, 1997). As Rapoport (2014) notes, back then Register was trying to apply environmental ideas that were already very popular at the time. Therefore, while the word *eco-city* only emerged in the 1980s, the concept of *eco-cities* 'originally emerged out of counterculture movements of the 1960s and 1970s as an approach to urban development that would respect environmental limits' (Rapoport, 2014: 137).

In his book, Register (1987: 3) defines an eco-city as an 'ecologically healthy city.' Ecology is central to the making of eco-cities. The *eco-city* concept incorporates the science of ecology to envision and ultimately produce urban spaces sensitive to their surrounding ecosystems. Register's argument is that ecology, with its scientific understanding of the complex and delicate relationships between organisms and their environment, can illuminate the negative impact of urbanization on the biotic and abiotic components of a given space. For Register (2006), the crux of the matter is that urban planning and urban design are often not informed by the laws of ecology, and therefore ignore ecosystems. In the theory of ecological urbanism, ecological ignorance is associated with ecological damage. When urban development is ecologically insensitive, the built environment, instead of growing in synergy with the natural environment, takes its place violently, thereby causing a loss of natural habitat and damaging ecosystem services.

In ecological urbanism, the blame falls particularly on modernism which is accused of viewing the city as an artefact designed and placed on a blank canvas, rather than on an intricate tapestry of ecological systems providing services, such as food, water and climate control, necessary to sustain human life. In Register's (1987) eyes, environmental problems are ultimately urban problems rooted in a conceptual misunderstanding of space typical of modernist urban planning. However, as Hall's (2002) studies show, there is not a single and universal modernist city. Modernist urban planning, from the end of the nineteenth century throughout the twentieth century, includes diverse ideologies and often contrasting ways of understanding cities and their making (Hall, ibid.; Lawton, 2020). It is therefore important to identify the specific modernist themes that ecological urbanism is reacting against, before delineating the position of ecological urbanists like Register. In this regard, there are two main lines of modernist thought which deserve attention. The first one is represented by Thomson's (1880) emblematic image of the *City of Dreadful Night* (Hall, 2002). This is in essence the city of the Industrial Revolution: dirty, polluted and darkened by the smokes of industry (see Figure 2.1). Its academic accounts tend to be Eurocentric and picture rapidly growing industrial cities such as London, Manchester and Glasgow, poorly planned and affected

FIGURE 2.1 The City of Dreadful Night captured in a sketch by Gustave Doré, later engraved by Adolphe François Pannemaker in 1872. It is always night in industrial London with its black smoky sky. The city's smoke-blackened slums look like a circle of Dante's hell and it is not a coincidence that Doré also produced some of the most iconic illustrations of *Inferno*.

Source: Wikimedia Commons

by slums and pollution (Benevolo, 1993; Gunn, 2000; Hall, 2002). This image of the city is unecological in the sense that natural spaces, abundant prior to the Industrial Revolution, are here being consumed to make room for factories (then commonly located inside cities) and to power industry, with severe repercussions in terms of environmental degradation and waste.

The second line of modernist thought that ecological urbanism seeks to counter is the Corbusian ideal city. Le Corbusier's urban ideals, conceived in the first half of the twentieth century, have had an incalculable impact on urban planning and their influence is still present in the making of contemporary cities (Hall, 2002). Le Corbusier's ideal city (1887–1965) is, graphically speaking, the antithesis of the City of Dreadful Night. While the industrial city of the late 1800s, as pictured above, appears dirty, chaotic and bereft of green spaces, Le Corbusier's model for the city of the early 1900s is the apotheosis of order with a shade of green. Coming from a family of watchmakers, the Swiss architect imagined a city where everything is planned in detail and controlled by the architect: a model of 'neat self-control, with not a blade of grass or a stray hair out of place' (Fishman, 1982; Hall, 2002: 219). The Corbusian city is hyper clean and tidy, and its core principles and philosophy become evident when looking at its visual representations in *Plan Voisin* (1925) and in the 1933 master plan for a new city *Ville radieuse* (*The Radiant City*). In addition to the radical geometrical order disciplining the built environment, what is noteworthy in Le Corbusier's master plans *à la* Voisin is the abundance of green areas, to the point that this image of the ideal city can be interpreted as a city of towers placed in a vast park (see Figure 2.2). However, the presence of greenery can be here conceptually misleading as it might hint at ideas of an ecological urbanism. This is not the case, because *green* and *ecological* are not necessarily synonymous with each other. As Le Corbusier (1929: 232) himself wrote in capital letters: 'WE MUST BUILD ON A CLEAR SITE!.' He believed in building cities from scratch, in order to start from a blank canvas, necessary in his mind to realize the perfect plans of the architect. This philosophy of city-making does not imply only the destruction of existing urban spaces (Le Corbusier demanded the demolition of Paris' city centre to implement his Plan Voisin), but also the annihilation of the existing natural environment. The green spaces of the Corbusian city are gardens and parks designed and built *ex novo*. They lie upon an ecological massacre systematically carried out to form the blank canvas desired by the architect, and exist exclusively in rigidly confined areas which are separated from the built environment.

Le Corbusier's urban ideals are symptomatic of a strand of modernist planning in which the natural environment is disconnected from the built environment. This is a conceptual fracture (and limitation) of modernism in which the *natural* and the *urban* are understood as oppositional and ultimately incompatible types of space which cannot exist in the same area. Akin to the work of Le Corbusier are the drawings of Ludwig Karl Hilberseimer (1885–1967) which take the Corbusian city to an extreme level of ecological insensitivity. In *Highrise city* (1924), for instance, Hilberseimer proposes a city where the natural environment has been completely

FIGURE 2.2 Le Corbusier's Plan Voisin (1925). Gargantuan towers of concrete and steel in a manicured garden were supposed to replace the Parisian city centre. Although to this day very influential in shaping the imagination of numerous architects and planners, this plan was never realized.

Source: Wikimedia Commons

replaced by the built environment, and nature finds no space at all (Figure 2.3). The common denominator in the City of Dreadful Night and the Corbusian ideal city is the so-called *domination of nature* thesis. This is the idea that humans can control natural forces and spaces, by means of science and technology (Leiss, 1994). In the City of Dreadful Night and the Corbusian city, the domination of nature thesis is interpreted in two different but interconnected manners. In the first case, *domination* stands for *violation*. The industrial city exploits the natural environment as fuel for industry and, in so doing, it permanently damages ecosystems. In the second case, *domination* stands for *subordination*. In the modernist city of the early 1900s, the natural environment belongs to a secondary class of space which is subject to the authority of the architect and confined within designed green areas. In both cases, nature is subjugated to human needs and severed from the built environment.

Every complex story has of course its exceptions and this is the case of modernism too. In the same periods discussed above, it is possible to find avant-garde thinkers who imagined then alternative forms of space sensitive to environmental concerns. The history of the eighteenth century, for example, is full of engineers and statisticians seeking to mitigate the pollution of the industrial city by modernizing its water

FIGURE 2.3 Ludwig Karl Hilberseimer's *Highrise city* (drawn in 1924, published in 1926) where the built environment reigns supreme.
Source: The Art Institute of Chicago

and waste infrastructures (De Block, 2016; Kaika and Swyngedouw, 2014; Rabinow, 1995). Specifically from an urban planning perspective, at the turn of the century Ebenezer Howard (1850–1928) was leading the Garden City movement to realize his vision of a city 'developed on the basis of its own natural resources, with total respect for the principles of ecological balance' (Hall, 2002: 8). These counter-current storylines demonstrate that the intellectual seeds of ecological urbanism can be found in an obscure past when cities were largely seen as unecological entities, and the term *eco-city* did not exist. The role of modernism is here ambivalent and deserves a twofold clarification. First, the *domination of nature* thesis, although prominent in modernism was inherited from Enlightenment philosophy (Harvey, 1989). Before then, dreams and methods of controlling natural forces were pushed forward by Francis Bacon in the Baroque era, and Bacon himself was ideologically influenced by Renaissance philosophers and early scientists (Leiss, 1994). Second, modernism, as a double-edged sword, pursued an often blind scientific and technological development responsible for the environmental horrors of the industrial city, but it also accelerated the development and refinement of numerous sciences, including the science of ecology. Modern ecology was established in the first half of the twentieth century and it is therefore a product of modernism. It later detached itself from modernist ideals when it embraced and drove the environment movements of the 1960s and 1970s but, as a science, ecology is intrinsically linked to modernism (see, for instance, Carson, 2000).

The environmentalism of the 1970s represents the cultural background of Richard Register and, thus, somehow paradoxically ecological urbanism rejects modernism while being profoundly indebted to it. The element of continuity is scientific and it lies in ecology which, as a modern science, should, so Register (2006) argues, inform urban planning in order to avoid repeating the environmental mistakes of the industrial city. The element of discontinuity is conceptual, and it lies in the modernist conceptual separation between the natural and the built environment, typical of the Corbusian city. Register's philosophy is openly against that urban ideal, and it seeks to break the divide between the *natural* and the *urban*, in the attempt to conceptualize spaces that are both natural and urban. His view of the urban is that of a hybrid environment in which nature and human societies compenetrate each other. In this sense, the philosophy of ecological urbanism resonates with Harvey's (1993: 31) claim that there is nothing unnatural about cities, and it is 'hard to see where "society" begins and "nature" ends' (see also Heynen et al., 2006; Gandy, 2003). What Register remarks is that such compenetration does not necessarily have to be conflictual. It can be harmonious, and ecology is the instrument to balance the expansion of human societies with the preservation of non-human organisms.

To plan an eco-city, Register reprises in urban terms the concept of *bioregionalism*. Advocated in the 1970s by environmentalists such as Peter Berg and Kirkpatrick Sale, bioregionalism is about recentreing the physical boundaries of human development, according to an ecological and geographical rationale (McGinnis, 1999). A bioregion, as a geographical area defined by the location of distinct geological features, the distribution of the regional flora and fauna, the presence of bodies of water and the manifestations of the local climate, is, for Register, more than a concept. It is a cartographic tool. In the theory of ecological urbanism, bioregions scientifically define the limits of urbanization and the boundaries of cities. Through the lens of bioregionalism, eco-cities emerge as urban settlements which do not cross a bioregion, instead growing within one, in sync with its ecosystems. An eco-city is then planned regionally, thinking beyond the area that is occupied by the built environment and considering all the spaces needed to sustain the metabolism of the city. This is a line of urban planning which actually predates 1970s environmentalism and can be traced back to the theories of Patrick Geddes (1854–1932), the father of regional planning and proposer of the idea of the *natural region*.

Geddes, a key figure in the complex story of modernist planning, took the concept of the natural region from *fin de siècle* French geography, arguing that the planning of a city should always begin with a survey of the natural resources of the region where that city sits (Geddes, 1905; Hall, 2002). Register's addition is ecology which, as a modern science, was not fully developed in Geddes's time. In eco-city planning, ecology is seen by Register as a precise scientific instrument which can ensure a harmonious urban coexistence between the human and the non-human. On a macro-scale, ecological studies can measure the *carrying capacity* of a bioregion, intended as the maximum population size that an area, given its natural resources and ecosystems, can sustain without collapsing (Register, 1987). On a micro-scale, ecology can calculate a *sustainable yield*: the amount of natural resources, such as timber for instance, which can be safely removed without upsetting a given ecosystem (Dasmann, 1985; Whitehead, 2007). In essence, ecology sets precise limits of urban growth, stopping urbanization when the resources consumed by a city are more than those produced by its bioregion.

However, the theory of ecological urbanism developed by Register reaches an impasse when the ideal concept of a bioregion inevitably clashes against the reality of urban growth. From a global perspective, as emphasized in Chapter 1, urbanization has been pushing the boundaries of cities further and further, regardless of the geology, geography and ecology of regions. Bioregions can have a precise perimeter and a rigid carrying capacity, but when a city grows horizontally and vertically, such limits are rarely taken into account by urban developers and stakeholders. This problem is addressed in another strand of ecological urbanism cultivated by Léon Krier, a social theorist, planner and architect representative of New Urbanism and New Classical Architecture: two movements openly against modernism and the *domination of nature* credo. Krier (1998) claims that modernist strategies of urban planning, fuelled by limitless processes of capital accumulation, have led to a monstrous urban growth responsible for the production of obese cities. Krier approaches urbanization from a metabolic perspective, in a way which is akin to contemporary studies on urban metabolism looking at the amount of natural resources consumed by cities to sustain their functions (Beloin-Saint-Pierre et al., 2017; Conke and Ferreira, 2015; Pincetl et al., 2012). In the work of Krier, urban obesity is a symptom of overexpansion, and manifests itself vertically and horizontally. In the first case, the centre of the city (its original nucleus) is contaminated by the proliferation of high-rise buildings, hyper density and the polarization of social and economic activities: a phenomenon which increases the value of land and the cost of rents. In the second case, suburbanization pushes urban growth outwards, toward lower costs of land, resulting in lower densities of buildings and activities. In both cases, such hypertrophy unbalances the structure, functions and aesthetics of the city, creating a gigantic and insatiable entity which devours the ecosystems upon and around which it sits.

For Krier (1998), the solution is urban growth by *duplication*, similar to Howard's (1898) original model for the Garden City, planned with fixed limits that, when reached, would trigger the development of a new one, eventually leading to 'a vast planned agglomeration' of Garden Cities (Hall, 2002: 94). Rather than expanding a city in a vertical and horizontal manner, Krier proposes to take the nucleus of an existing city and replicate it in a different geographical space. A mature urban settlement would then stop growing before becoming obese, instead becoming the matrix for the construction of new urban settlements. On these terms, ecological urbanism can be explained through the concept of *gemmation*. In biology, gemmation, or budding, is an asexual form of reproduction by gemmae. It produces a bud which, when mature, detaches itself as a genetically identical but smaller version of its parent (Huh et al., 2003; Soifer et al., 2016). The process of gemmation can be found in cellular reproduction, animal reproduction and plant multiplication. In all these cases, the new cell or the new organism is a copy of the mother cell or the parent organism. From an urban perspective, gemmation would produce a city, smaller than its matrix, characterized by a sustainable metabolic demand, and with a growth potential meant to accommodate a growing population. Bearing the traits of its parent, the new city would start its life with the same qualities, services and shape of a city centre: a matter of urban design which will be explored in the next section.

Designing the eco-city

The design of an eco-city is a topic which is at the core of Register's theory of ecological urbanism. Ecology is again central and, in relation to urban design, it assumes two meanings. First, ecology as a science is applied to protect the ecosystems upon and around which a city is placed. In this case, the rationale mirrors the planning theory discussed in the previous section. For Register (1987), cities need to be designed in a way which is sensitive to local ecosystems. This implies the minimization of the built environment in order to favour the preservation of the natural environment. Register observes how in contemporary cities, a considerable portion of the urban fabric is reserved for cars. Highways, road junctions, vehicle lanes and parking spaces are materially occupying areas which could otherwise be designed as green spaces. This is only the beginning. According to Register (ibid.), the concept of green spaces has to be deconstructed through the science of ecology, by following what he calls a *green hierarchy*. In this type of urbanism, a hierarchical approach to the ecology of a city would prioritize the conservation of all the plants that can be used as sources of food and energy, as well as for pharmaceutical purposes. Green spaces with sole ornamental function would be disregarded. Landscape architecture, in the shape of lawns and purely recreational gardens, for example, needs maintenance and considerable investments in terms of water, energy, time, fertilizers and, above all, space which could be used to bring agriculture inside the city.

Reducing spaces for cars requires the reduction of cars which, in turn, requires the reduction of the need for cars. To tackle this issue, Register draws upon traditional and vernacular ideas of urbanism, predating modernism and the automobile era, and depicts the eco-city as a compact city. For him, eco-cities are *three-dimensional cities* rather than *flat cities*, and adhere to *inner development* rather than *outer development*. His vision is that of a city whose inhabitants do not need to travel long distances to access urban services. This line of thinking has a long tradition in urban design, and Register refers explicitly to Medieval European cities confined within a walled perimeter and largely pedestrian. There are of course many other examples, not mentioned by Register, which resonate with the theory of ecological urbanism. Aleppo (Syria) and Shibam (Yemen), for instance, are some of the most prominent cases of compact urban design in the Middle East. The genesis of these settlements belongs to a time when the mobility of people was considerably limited by their capacity to walk, under challenging weather conditions which could not be mitigated by the air-conditioning of a modern means of transport. These cities were therefore designed in a compact manner, not only to bring people and services closer to each other, but also to generate natural shading, channel the winds of the desert and ultimately decrease the perceived temperature in the built environment.

Although Register considers compact urban design only in relation to cities of a modest size and to towns, it has to be noted that the same principles can, to some extent, be applied to larger urban settlements. Fifteenth-century Aleppo, for example, notable for its compact urbanism, was not a small town (Hourani, 2013). It was an important economic centre linking different agricultural districts and trade routes. It has been estimated that, during that epoch, the local population was around 80,000 people: a number which does not include the thousands of merchants, farmers and buyers from all around Syria, who regularly frequented its streets and markets (Hourani, ibid.). Aleppo kept growing, maintaining for centuries its compact design (see Figure 2.4). Dubai itself, now the quintessence of flat urbanism and unsustainability, used to be part of this category. In the pre-oil era, Dubai was a compact settlement characterized by narrow crooked alleys, whose spirit can be still felt today by walking in the historic district of Al Bastakiya.

Within the compact city framework, Register continues to employ ecology in the design of the eco-city, this time, however not as a science, but as a metaphor for social diversity. In this sense, ecology is not only a scientific instrument which

FIGURE 2.4 Aleppo in 1697, portrayed by Henry Maundrell (1665–1701).
Source: Historic Cities, the Hebrew University of Jerusalem and the Jewish National and University Library

the urbanist can use to protect the biodiversity of a region, but also a source of inspiration to achieve and preserve diversity with regard to services, infrastructures and ultimately citizens. For Register (1987), urban diversity can be obtained by means of *integral neighbourhoods* which mix different activities, services and infrastructures. In this urbanism, thinking in terms of diversity implies thinking about what people need (shelter, food, healthcare and education, for instance) and like (sport, music and art, for example), and then connecting those needs and desires to the infrastructures that can support them (homes, markets, hospitals, schools, gyms, theatres, galleries, etc.). From a philosophical perspective, there are echoes of Aristotle's *eudaimonia* introduced in Chapter 1. Urban spaces and infrastructures are mediums for people to discover and nurture their own potential, in order to eventually reach happiness. This vision of the city privileges density to bring a wide range of activities and related infrastructures within the same neighbourhood, in close proximity to each other. Inside this dense multifunctional environment, citizens would not have to commute, and everything needed and wanted, would be relatively close to their home.

In ecological urbanism, the diversity of urban spaces, and of the social activities that they incarnate, reflects the diversity of the people who populate them. Urban design brings people closer to each other not just physically to reduce the need for cars, but also culturally and socially to cultivate a diverse and, yet, harmonious society. Register (1987) posits that as a healthy ecosystem is one characterized by rich biodiversity, a healthy city is one characterized by rich social diversity. A compact city with integral neighbourhoods can, according to the theory of ecological urbanism, support the convergence of people, thereby avoiding social isolation and fractures. This is a theory which evokes recent debates in human geography and urban sociology, over the socio-cultural power of *encounters* (see Darling and Wilson, 2016; Wilson, 2017). The argument is that an urban encounter with a stranger, an acquaintance or a friend, can create and strengthen social relationships in a city, thus providing the baseline for the formation and preservation of its community (Valentine, 2008). Going back to Aristotle's (1992) conceptualization of the city as a social entity, encounters would then animate a fundamental aspect of the urban, without which cities would collapse. Although a single encounter might be almost weightless, a multitude of encounters have the potential to connect many individuals to other inhabitants of the same environment: an action which, as postulated in Allport's (1954) *contact hypothesis*, serves an important educational function. The premise is that people are uncomfortable with the unknown. Therefore, contact between different groups can reduce the feeling of uncertainty, eventually producing a sense of knowledge and familiarity among strangers, with powerful effects against social plagues such as marginalization and racism.

A similar social aspect of urban design is celebrated in the work of Jane Jacobs. Taking a conceptual stance akin to Register's theories of urban design, Jacobs (1993: 19) stresses 'the need of cities for a most intricate and close-grained diversity of uses,' arguing that 'the art of city design, in real life for real cities, must become the science and art of catalysing and nourishing these close-grained working rela-tionships.' She questions the design of cities in relation to different components and scales, putting emphasis on the accessibility of sidewalks, streets, parks and neigh-bourhoods: every urban space, no matter its scale, matters. For her, urban spaces need to be open and accessible from multiple directions, for diverse users and via several entries, to maximize the circulation of people. Moreover, the built envir-onment should offer heterogeneous activities which, in turn, would generate a semi-permanent gravitational pull over various individuals 'who go outdoors on different schedules and are in the place for different purposes' (Jacobs, 1993: 198). Like in Register's urbanism, the city envisioned by Jacobs would be invigorated by a constant pulse of social activities bringing people together and *eyes upon the street* throughout the day. The eyes of those who live, know and protect the city, just by being present in space. This would be a city of mutual understandings, rather than racial discrimination. A vigilant and hence safe city, rather than a deserted envir-onment prone to crime.

Seen from this perspective, with urban ecology symbolizing social diversity, ecological urbanism reveals a social dimension which is not often acknowledged. This is a common fallacy since the word *eco-city*, with its prefix *eco*, immediately summons images of a green city and of luxuriant urban landscapes covered with plants of every kind. In reality, the *ecological* is only half of the eco-city narrative. Ecological urbanism, theoretically, deals with a large part of the spectrum of the urban. The city is understood as a physical entity with a physical and potentially catastrophic impact on the natural environment, in dire need of ecological lessons. However, the city is also a place of complex social relationships: a dimension which can too be developed in ecological terms. Ultimately, ecology is the science of equilibrium. Regardless of the particular species and geographical spaces that are taken into account, ecology is about balancing the interactions which occur in a given environment. For this reason, an ecological urbanism cannot ignore the mechanics of human societies, and must address the way people interact in the city and how they interact with the city itself. For the same reason then, the theory of ecological urbanism cannot but also be a political theory, with an emphasis on the governance of urban settlements.

Governing the eco-city

The political dimension of ecological urbanism can be traced back to several phi-losophical currents, often in contrast with each other. While it is evident that the development and management of an eco-city requires politics (like in every city), there is no universal agreement on what political system can best achieve and sus-tain the ecological and social diversity discussed in the previous sections. Register's theory of the eco-city, for instance, resonates with the concept of democracy which is a recurring but never explicit theme in *Ecocity Berkeley*. Register (1987: 49) affirms that the process of developing an eco-city should include as many people as possible: 'it is not vicarious but participatory, not to be dictated, but to be created in a million ways simultaneously from the grassroots to the highest levels of planning and back down again, with a role for each of us.' This line of democratic thinking is connected to the idea of designing cities which maximize the diversity of services, infrastructures and users. Given that an eco-city should cultivate all the different needs, dreams and desires of its citizens, the act of eco-city-making should be constantly informed by an understanding of what these needs, dreams and desires are. Hence, for Register (ibid.), it is of paramount importance, to democratically include all members of society in eco-city initiatives, for them to express their vision of the city.

Register is also aware that the issue of democracy is connected to the issue of education. His argument is that the extent to which people are willing to work together towards the development of eco-cities, depends on their knowledge and cultural background, particularly when it comes to ideas of ecology and the functioning and functions of ecosystem services. He (1987: 51) contends that 'many people simply haven't had sufficient positive exposure to such ideas,' and believes that comprehending the ecological rules which regulate the natural environment, inevitably leads to their appreciation and observance. On these terms, ecological urbanism echoes the philosophy of Spinoza (1996) according to which mastering the knowledge of reality and grasping the structure of the universe matures a love for the universe itself. For Spinoza (ibid.), an individual's ethical conduct in relation to something is dictated by the individual's level of knowledge about the subject matter. However, the Dutch philosopher takes a firm stand, arguing that humans are intrinsically selfish and tend to operate in irrational manners, hence the need to regulate their behaviour (Spinoza, 2017). These are extremely complex philosophical (as well as practical) questions of governance. Spinoza himself acknowledges the danger of a quick fix through a top-down draconian governance, inasmuch as authoritarian and despotic regimes would not produce citizens but rather slaves. To address these questions, the theories of urban planning and design developed by urbanists like Register and Krier fall short, and ecological urbanism needs to enter the realms of political philosophy and political science. More specifically, the political theory concerning eco-cities threads a line with environmental philosophies which deal explicitly with the connection between ecology and politics in spatial terms. Following this thread, the conceptual seeds of the eco-city can be found in social ecology and eco-socialism whose philosophical systems will be elucidated in the remainder of this section.

The philosophy of social ecology is largely derived from the work of Murray Bookchin (1921–2006), an American political theorist of Russian origins. At the centre of Bookchin's philosophical system is the idea that the social realm and the ecological realm are intrinsically interconnected. Building upon this premise, the aim of his philosophy is to theorize and encourage an equilibrium between social and ecological needs (Pepper, 1996; Whitehead, 2007). As Roseland (1997: 199) points out, for Bookchin the key issue is 'not simply protecting nature,' but rather creating a society which is 'in harmony with nature.' Given that in social ecology, the *ecological* is only half of the equation, focusing solely on the preservation of the natural environment is seen as a practise which is as dangerous as the exclusive development of human societies. This is a radical line of environmental and political thinking which goes against a contemporary of Bookchin, Arne Naess whose philosophy of *deep ecology* is based on the substitution of anthropocentrism with ecocentrism (Naess, 2008). Bookchin is openly against any form of centrism and hierarchy, and embraces anarchism as the only type of political organization capable of eliminating, at the same time, the social and environmental problems of modern societies.

In the words of Bookchin (2007: 19), 'nearly all of our present ecological problems originate in deep-seated social problems.' This hypothesis derives from the assumption that 'the notion that man must dominate nature emerges directly from the domination of man by man' (Bookchin, 1971: 63). Hence the urgent need of addressing political questions first, to subsequently fix related socio-environmental issues. Bookchin argues that humans dominate each other and, consequently, dominate nature, because of the 'coercive power of the state,' claiming that 'without the state and other structures of exploitation associated with hierarchy among people, environmental problems would not arise' (Adams, 2003: 186). This claim is based on the understanding of the state as a political entity which produces and reproduces a condition of *hierarchy* and *gigantism*. More specifically, in this condition, single individuals find themselves, first, within a hierarchical structure in which they give a substantial portion of their agency to a minority of people who control political power. Second, individuals become part of a gigantic and endlessly growing political entity, the state, whose sheer size is incomparable with that of the individual. Inside this leviathan, the individual ends up being a microscopic and almost insignificant cogwheel.

For a social ecologist like Bookchin, this socio-political situation is ecologically problematic because the state keeps growing without limits and, in so doing, it consumes an ever-increasing amount of resources, thus polluting the planet. Most importantly, and here lies the core of social ecology, within gigantic political entities, the individual is detached from production inasmuch as, geographically, industry is located in a space which is not where the individual lives. For this geographical reason, individuals are not exposed to the negative environmental outcomes of industry, and do not perceive and, more in general, experience pollution and waste. The environmental horrors of industrialism are elsewhere. Far away from the eyes and conscience of people. Therefore, this lack of experience makes the individual insensitive to ecological problems, and less inclined to react to them. Furthermore, when an individual is sensitive to environmental issues and would like to change the state of affairs, he or she cannot do so, because the hierarchical system denounced by Bookchin has taken much of his or her agency.

According to the American-Russian philosopher and anarchist, ecological issues 'cannot be resolved within the existing social framework' (Bookchin, 1978: 23). The state should be eliminated and replaced by small-scale communities which autoregulate themselves. This is a form of decentralized communitarianism driven by a strong belief in egalitarianism, and characterized by an inclusive process of decision-making (Harvey, 1996). In a social ecologist community, 'production is where people live,' and everything happens within close boundaries (Pepper, 2002: 178). People cannot avoid facing environmental problems. The individual has therefore a strong double incentive to live in a way that does not damage the local ecology, and to promote a collective development which limits pollution and waste. Moreover, within this political context free of hierarchy, individuals maintain their agency and can act to integrate the principles of ecology into their community. From an urban perspective, in the philosophy of social ecology, large cities disappear and are substituted by a constellation of small settlements, emerging from a sea of natural environments (Bookchin, 1992). Due to the small scale of the urban, there is no room for externalization. Politics cannot be externalized, in the sense that everyone participates in the political life of the place. Moreover, given the proximity of factories and, more generally, means of production, the negative by-products of industry, waste and pollution, cannot be externalized either, and production is therefore consciously kept down to a minimum.

The second environmental philosophy pertaining to the governance of eco-cities, eco-socialism, takes a less radical approach. Instead of professing the dissolution of the state, eco-socialists look at how the architecture of mainstream political systems can be reconfigured. Here the conceptual foundations are Marxist theories of social change. Like social ecology, eco-socialism sees environmental and social issues as interlinked, and approaches socio-environmental problems through the lens of Marxism (Borgnäs et al., 2015; Kovel, 2019; Huan, 2014). Before the advent of eco-socialists such as Rudolf Bahro (1935–1997) and Joel Kovel (1936–2018), Marx and Engels had already denounced the environmental damages which social distortions could unevenly inflict upon society. In the first volume of *Capital*, for example, Marx (2004) laments the harm that factories' pollution, noise and heat can cause to the human body. In *The conditions of the working class in England*, Engels (1987) is horrified by the urban spaces produced by the Industrial Revolution, and describes in detail the filthy and unhealthy working-class neighbourhoods of the industrial city. Building upon Marxism, the philosophy of eco-socialism identifies in capitalism the mother of all socio-environmental problems, and focuses the analysis on the mechanics of capitalist political economies, in order to understand how they interfere with social and ecological systems (Pepper, 2002). The rationale is that by understanding how capitalism functions, it would also be possible to understand how capitalism can be dismantled (Johnston, 1989).

In eco-socialism, capitalism is held responsible for two main problems. First, a capitalist economy requires constant production and expanding markets in order to generate profit. The issue lies in the fact that constant production depends upon the constant extraction of resources and industrialization which, together, produce environmental degradation and pollution. Second, capitalism relies on inequality and reproduces inequality. In this sense, a classic Marxian critique of capitalist systems is that when capitalists sell a product, the exchange value of that product is greater than what was paid for labour power. Ultimately, this dynamic generates asymmetry in terms of wealth and power relationships, since the capitalist is earning more than the labourer. When this social problem inevitably meets the environmental problem discussed above, from a geographical and urban perspective, the result is a minority of wealthy people who can shield themselves from environmental bads because, for instance, they can afford to live in the greenest and cleanest parts of the city: what Hodson and Marvin (2010) would call a *premium ecological enclave*. The urban poor cannot, and theirs is the space that is environmentally degraded and polluted the most.

For eco-socialists, the solution to both social and environmental problems is the abolition of capitalism, and the establishment of socialism as a potential bridge to communism. The argument advanced is that in an eco-socialist political system, since capital and the prospect of profit do not exist, there is less pressure for production and, therefore, less extraction of natural resources and industrialization. Moreover, this is a society where classes have been abolished and there is a symmetry of wealth. Within such a system, according to eco-socialist theories, people can afford the same products and services. Following this line of philosophical thought, in geographical and urban terms, there would be a common interest in reducing pollution and respect ecosystems. Seeing that everybody is, figuratively speaking, in the same boat, no rich minority could pay its way out of environmental problems. The environmental burdens and benefits of the city would be equally distributed, and a condition of environmental justice would be achieved.

It is important to emphasize that, in the context of ecological urbanism, both social ecology and eco-socialism should not be intended as rigid blueprints. Traditionally, the purpose of philosophy is not to dictate development in a step-by-step way with specific and unavoidable procedures, but rather to shed light, provoke, inspire and, at times, denounce. On these terms, political philosophy (and its environmental and urban strands in particular) is not meant to provide an agenda requiring the signature of politicians, policy-makers, urbanists, architects and citizens. What the political theories discussed in this section critique is the role of capitalism and, more specifically, its incapacity to fix or simply mitigate the many socio-environmental problems which affect human societies. When it comes to the development of eco-cities then, taking into account the theories of eco-socialists and social ecologists implies reflecting not only on the physical structure of a city but also on the structure of its political system and on the political economy which that system promotes. With this line of thinking, governance becomes as necessary as urban planning and urban design for the realization of an ecological urbanism. Thus, any existing eco-city initiative must be examined and evaluated against its overarching political system: a perspective which will be central in the second part of the book.

Conclusions

The concept of ecological urbanism is not self-evident. Far from being axiomatic, it has ideological foundations and origins whose excavation requires conceptual tools from a variety of disciplines. As a result, the meaning of *eco-cities* often relies on subjective interpretations and, therefore, as the second part of the book will show, it is prone to manipulation. This chapter has noted that there are however cardinal themes and aspects, bearers of specific meanings, which should not be bypassed when the term *eco-city* is employed, whether theoretically or practically. Ignoring them means that the subject matter, being it a vision or an actual project, is not an eco-city. The moment the many theories of ecological urbanism are assembled, the figure that appears might not be completely clear, but the message that it emanates is loud and significant. The eco-city is science and philosophy. It is underpinned by notions of urban ecology whose precise methodologies can be used to plan and design urban spaces which are in sync with local and regional ecosystems. In this sense, the eco-city is a pragmatic concept followed by equally pragmatic concepts, such as *sustainable yields* and the *carrying capacity* of a bioregion, which urbanists can translate into ecologically sensitive urban agendas. Additionally, the eco-city has a strong philosophical dimension. It shines a conceptual light which makes urbanists see the city as a space of social diversity, where being different is optimal rather than problematic. On these terms, the eco-city does not envision cities, but rather societies. It is a theory of social and political organization in which the built environment becomes a medium of socio-political change.

Moreover, the theory of ecological urbanism does not conceive technology and technological innovation as fundamental aspects of urban development. Instead, it draws upon ideas of vernacular urbanism which predate the Industrial Revolution and the cult of modernity. Eco-cities are ideologically connected to ancient cities which proved to be sustainable, without the support of high-tech infrastructures. The story of Aleppo and Shibam, for instance, shows that environmental and social sustainability can, to some extent, be addressed simply through urban design. Natural shading, ventilation, mobility and encounters can be realized and maximized, by looking backwards rather than forwards. Following this line of thought, finally, the eco-city emerges as a radical and counter-current system of thinking. It denounces those strands of modernism obsessed with technological development and blind to its environmental repercussions. It critiques cities' dependence on technology, urging urban designers and planners to liberate the built environment from the chains of the car. Above all, it condemns capitalist political economies for reproducing a system of endless industrialism whose machinery is hidden to the individual, but whose environmental horrors are well known to the urban poor. As an idea, the eco-city comes from different sources which individually are incomplete and flawed, but together they send a powerful message which is simultaneously *ecological, social* and *political*. When only part of this message is heard, sustainability will not be achieved.

References

Adams, W. M. (2003). *Green Development: Environment and sustainability in the Third World*. Routledge, London.
Aristotle (1992). *The Politics*. Penguin Classics, London.

Allport, G. W. (1954). *The nature of prejudice*. Basic Books, New York.

Beloin-Saint-Pierre, D., Rugani, B., Lasvaux, S., Mailhac, A., Popovici, E., Sibiude, G., Benedetto, E. and Schiopu, N. (2017). A review of urban metabolism studies to identify key methodological choices for future harmonization and implementation. *Journal of Cleaner Production*, 163, S223–S240.

Benevolo, L. (1993). *The European city*. Blackwell, Oxford.

Bookchin, M. (1971). *Post-scarcity anarchism*. Ramparts Press, Berkeley.

Bookchin, M. (1978). Ecology and revolutionary thought. *Antipode*, 10 (3–1), 21–21.

Bookchin, M. (1992). *Urbanization without cities: The rise and decline of citizenship*. Black Rose Books Ltd., Montreal.

Bookchin, M. (2007). *Social ecology and communalism*. AK Press, Oakland.

Borgnäs, K., Eskelinen, T., Perkiö, J. and Warlenius, R. (Eds.). (2015). *The politics of ecosocialism: Transforming welfare*. Routledge, London.

Caprotti, F. (2014). *Eco-cities and the transition to low carbon economies*. Springer, Berlin.

Carson, R. (2000). *Silent spring*. Penguin Modern Classics, London.

Chang, I. C. C. and Sheppard, E. (2013). China's eco-cities as variegated urban sustainability: Dongtan eco-city and Chongming eco-island. *Journal of Urban Technology*, 20 (1), 57–75.

Conke, L. S. and Ferreira, T. L. (2015). Urban metabolism: Measuring the city's contribution to sustainable development. *Environmental Pollution*, 202, 146–152.

Cugurullo, F. (2016). Urban eco-modernisation and the policy context of new eco-city projects: Where Masdar City fails and why. *Urban Studies*, 53 (11), 2417–2433.

Darling, J. and Wilson, H. F. (Eds.). (2016). *Encountering the city: Urban encounters from Accra to New York*. Routledge, London.

Dasmann, R. F. (1985). Achieving the sustainable use of species and ecosystems. *Landscape Planning*, 12 (3), 211–219.

De Block, G. (2016). Ecological infrastructure in a critical-historical perspective: From engineering 'social' territory to encoding 'natural' topography. *Environment and Planning A*, 48 (2), 367–390.

Engels, F. (1987). *The condition of the working class in England*. Penguin, London.

Fishman, R. (1982). *Urban utopias in the twentieth century: Ebenezer Howard, Frank Lloyd Wright, and Le Corbusier*. MIT Press, Cambridge.

Gandy, M. (2003). *Concrete and clay: reworking nature in New York City*. MIT Press, Cambridge.

Gandy, M. (2015). From urban ecology to ecological urbanism: an ambiguous trajectory. *Area*, 47 (2), 150–154.

Geddes, P. (1905). Civics: as Applied Sociology. *Sociological Papers*, 1, 101–144.

Gunn, S. (2000). *The public culture of the Victorian middle class: ritual and authority and the English industrial city, 1840–1914*. Manchester University Press, Manchester.

Hagan, S. (2014). *Ecological urbanism: the nature of the city*. Routledge, London.

Hall, P. (2002). *Cities of tomorrow*. Blackwell, Oxford.

Harvey, D. (1989). *The condition of postmodernity*. Blackwell, Oxford.

Harvey, D. (1993) The nature of environment: dialectics of social and environmental change. In Miliband, R. and Panitch, L. (Eds.). *Real problems, false solutions. A special issue of the Socialist Register*. The Merlin Press, London.

Harvey, D., (1996). *Justice, nature and the geography of difference*. Blackwell, Oxford.

Heynen, N. C, Kaika, M. and Swyngedouw, E., (Eds.). (2006). *In the nature of cities: urban political ecology and the politics of urban metabolism*. Routledge, London.

Hodson, M. and Marvin, S. (2010). Urbanism in the Anthropocene: Ecological urbanism or premium ecological enclaves?. *City*, 14 (3), 298–313.

Hourani, A. (2013). *A history of the Arab peoples: Updated edition.* Faber & Faber, London.

Howard, E. (1898). *To-morrow: A peaceful path to real reform.* Swan Sonnenschein, London.

Huan, Q. (Ed.). (2014). *Eco-socialism as politics: Rebuilding the basis of our modern civilisation.* Springer, Berlin.

Huh, W. K., Falvo, J. V., Gerke, L. C., Carroll, A. S., Howson, R. W., Weissman, J. S. and O'shea, E. K. (2003). Global analysis of protein localization in budding yeast. *Nature,* 425 (6959), 686.

Jacobs, J. (1993). *The death and life of great American cities.* The Modern Library, New York.

Johnston, R. J. (1989). *Environmental problems: nature, economy and state.* Belhaven Press, London.

Kaika, M. and Swyngedouw, E. (2014). Radical urban political-ecological imaginaries: Planetary urbanization and politicizing nature. *Dérive,* 55, 15–20.

Krier, L. (1998). *Architecture: Choice or fate.* Papadakis Publisher, Singapore.

Kovel, J. (2019). *The emergence of ecosocialism.* 2Leaf Press, New York.

Lawton, P. (2020). Tracing the provenance of urbanist ideals: A critical analysis of the Quito papers. *International Journal of Urban and Regional Research.* doi:10.1111/1468–2427.12871.

Le Corbusier (1929). *The City of Tomorrow and its Planning.* John Rodher, London.

Leiss, W. (1994). *Domination of nature.* McGill-Queen's University Press, Montreal.

McGinnis, M. V. (Ed.). (1999). *Bioregionalism.* Routledge, London.

Marx, K. (2004). *Capital,* volume I. Penguin, London.

Mostafavi, M. and Doherty, G. (Eds.). (2016). *Ecological urbanism.* Lars Müller, Zurich.

Naess, A., (2008). *Ecology of wisdom: Writings by Arne.* Penguin Books, London.

Pepper, D. (1996). *Modern environmentalism: An introduction.* Routledge, London.

Pepper, D. (2002). *Eco-socialism: From deep ecology to social justice.* Routledge, London.

Pincetl, S., Bunje, P. and Holmes, T. (2012). An expanded urban metabolism method: Toward a systems approach for assessing urban energy processes and causes. *Landscape and Urban Planning,* 107 (3), 193–202.

Rabinow, P. (1995). *French modern: Norms and forms of the social environment.* University of Chicago Press, Chicago and London.

Rapoport, E. (2014). Utopian visions and real estate dreams: the eco-city past, present and future. *Geography Compass,* 8 (2), 137–149.

Register, R. (1987). *Ecocity Berkeley: building cities for a healthy future.* North Atlantic Books, Berkeley.

Register, R. (2006). *Eco-cities: Rebuilding cities in balance with nature.* New Society Publishers, Gabriola Island.

Roseland, M. (1997). Dimensions of the eco-city. *Cities,* 14 (4), 197–202.

Soifer, I., Robert, L. and Amir, A. (2016). Single-cell analysis of growth in budding yeast and bacteria reveals a common size regulation strategy. *Current Biology,* 26 (3), 356–361.

Spinoza, B. D. (1996). *Ethics.* Penguin Books, London.

Spinoza, B. D. (2017). *Political treatise.* Hackett, Indianapolis.

Thomson, J. (1880). *The city of dreadful night, and other poems.* Reeves and Turner, London.

Valentine, G. (2008). Living with difference: reflections on geographies of encounter. *Progress in human geography,* 32 (3), 323–337.

Whitehead, M. (2007). *Spaces of sustainability: geographical perspectives on the sustainable society.* Routledge, London.

Wilson, H. F. (2017). On the paradox of 'organised' encounter. *Journal of Intercultural Studies,* 38 (6), 606–620.

3

THEORIES OF SMART URBANISM

Introduction

In the twenty-first century, the smart city has become a dominant urban ideal framing how the majority of urban developers and policy-makers think about the development of new and existing cities. It escapes a geographical concentration, spreading its influence across the world (Karvonen et al., 2018a). Stories of smart urbanism are told across different spaces and scales, influencing the perception of what is deemed desirable for the present and future of the city (Bina et al., 2020; Burns et al., 2021; Joss et al., 2019). However, a single universal definition and image of the smart city has not emerged in the literature. While smart-city initiatives abound, their meaning is elusive and their vision nebulous (Carvalho, 2014; Glasmeier and Christopherson, 2015; Haarstad, 2017). Yet, within this complex constellation of projects for smart cities, it is possible to find commonalities in terms of shared characteristics and recurring tensions, in order to make sense of the meaning and implications of being *smart* in an urban context. First and foremost, smart urbanism is an urbanism which is typically enabled by Information and Communication Technology (ICT). In this sense, prominent is the production of data or, simply put, information about the mechanics and functioning of key aspects of the city, such as transport, energy (production, circulation and consumption), security and governance (Albino et al., 2015; Caragliu et al., 2011; Coletta et al., 2018; Cugurullo, 2020; Silva et al., 2018). As Kitchin (2014, 2019) and Batty (2013, 2016) emphasize, the data generated via ICT and then processed by so-called smart cities is *big* because of the sheer volume of the information that is gathered, and *real-time* due to its constant updating and instantaneous transferability (see also Kitchin and McArdle, 2016; Leszczynski, 2016). The data is also *actionable*, inasmuch as policy-makers working in smart cities have the possibility to use it and act upon it, for the purpose of making informed decisions over the development and management of a given urban space.

Second, such urbanism tends to be premised on the idea of the city as a chaotic, dysfunctional and ultimately inefficient system. On these terms, smart urbanism with its actionable, real-time big data, is hailed as a bringer of rationality and order, meant to 'tame the unruly city' (Allwinkle and Cruickshank, 2011; Goodspeed, 2015; Karvonen et al., 2018b: 3). Building upon a storyline in which an inefficient city is an unsustainable city, the notion of smart urbanism has been increasingly associated with the notion of sustainable urbanism. As a result, smart-city discourses today frequently overlap with discourses on sustainability (Angelidou et al., 2018; Bibri and Krogstie, 2017; Höjer and Wangel, 2015). For instance, common is the assumption that increasing the energy efficiency of cities through smart grids coupled with renewable energy technologies, can decrease energy waste, reduce urban carbon emissions and thus mitigate climate change. According to this narrative, *smart* seems to have become the new *sustainable* (Martin et al., 2018). In relation to the now widespread *smart* and *sustainable* syncretism, the third significant aspect of the smart-city phenomenon is that these discourses and narratives are rarely tailored around specific spaces and times. Instead, they are universal in nature and scope. Smart urbanism, as mass media and stakeholders often portray it, is supposed to be applicable everywhere in the world, as an urban model that can be superimposed on any given place. Because of such a connotation, it has a universalizing impetus set to pass through local contextual barriers, thereby replacing vernacular urbanisms.

Fourth, the presumed universality of smart urbanism stems from the understanding of the smart city as an apolitical artefact which can be implemented regardless of local political orientations, cultures and social dynamics (Shelton et al., 2015). This aspect, in turn, stems from the main enablers and promoters of smart urbanism: ICT companies. Smart-city initiatives are largely shaped by private companies like IBM, Cisco and Siemens, which provide the many information and communication technologies meant to enable a smart urbanism (Barns et al., 2017; McNeill, 2015). There is a twofold implicit promise here. The technology in question is neutral (meaning that political, social and geographical differences do not affect its performance) and it can transform every ordinary city into a smart city. Of course, the expected transformation is not free. Since smart-city solutions are provided by the private sector, they come with bills which are not cheap. Smart urban technologies tend to be expensive and usually need constant updating and maintenance. For these reasons, smart urbanism is a capital-intensive urbanism requiring considerable investments from local municipalities and national governments. Finally, a defining characteristic of smart urbanism is its presumed existence. Especially among urban developers and practitioners, the smart city is rarely discussed as a theory, but rather as a matter of fact. By looking at the myriad of smart-city brands and logos attached to cities from around the world, and observing the prevalent utilization of ICT in the management of urban infrastructures and services, smart cities appear to be a diffused reality. Seen from this perspective, the smart city is everywhere and in a way 'we are already living in the smart city age' (Kitchin et al., 2018: 1; Mora and Deakin, 2019).

However, critical academic interventions, particularly from the fields of urban geography, urban planning and science and technology studies, have stressed that the above key characteristics of smart urbanism must be taken into account together with a series of inner tensions and contradictions (Luque-Ayala and Marvin, 2015; Marvin et al., 2015). Capturing such intrinsic issues is essential in order to fully understand the complexity of the smart-city phenomenon. First, although smart-city initiatives seem to be taking place in almost every city, in reality as Caprotti (2019: 2466) remarks, 'there are few places and spaces in the contemporary city that can be visualized and made legible as clearly belonging to the smart city.' What a smart city is, does and looks like is not evident. Second, smart-city projects do not have a standard materiality. The lens of geography has highlighted that smart urbanism has not one, but a plethora of different contextual incarnations. The smart-city phenomenon has been interpreted in geographical studies, as a context-dependent phenomenon which 'does not occur in a vacuum' and is ultimately doing 'different work in different places' (Dameri et al. 2019; Karvonen et al., 2018b: 4; Kong and Woods, 2018; McFarlane and Söderström, 2017). The *smart city* concept is filtered and processed through specific political economies and cultures, thereby leading to the formation of peculiar rather than universal urban environments. Hence, it would be more appropriate to talk about smart *urbanisms*, plural, as opposed to smart urbanism, singular.

Third, it is also not evident if a smart-city logo corresponds to a progressive plan of urban development set to improve the efficiency of the built environment, or if it masks a charade. Similar controversies have been denounced many times, in relation to projects promoted under the shadow of the 'sustainability' banner, to hide what in fact were traditional pro-growth economic agendas (Imrie and Lees, 2014; Raco, 2005). What appears to be clear from empirical studies of smart-city initiatives is that smart urbanism does not necessarily bring order. Despite being commonly associated with images of rationality and discipline, smart urbanism can actually bring chaos, conflict and further fragment already splintered cities (Cowley et al., 2018; Cugurullo, 2018a; Graham and Marvin, 2002; March and Ribera-Fumaz, 2016). Fourth, smart urbanism has been repeatedly critiqued as being the ICT incarnation of neoliberal urbanism. A number of scholars have seen in the invasion of large ICT companies in the planning and management of the city, a stepping back of the state to the detriment of non-monetizable public urban issues (Cardullo and Kitchin, 2019; Glasmeier and Nebiolo, 2016; Grossi and Pianezzi, 2017; Trencher, 2019). Following this argument, smart urbanism has been accused of hindering a holistic and sustainable development in favour of business priorities. Smart interventions have often pushed for the development of only a minority of spaces and segments of the population, on the basis of a very narrow interpretation of sustainability which misses a balance among different social, environmental and economic spheres (Colding and Barthel, 2017; Kaika, 2017; Lee et al., 2020; Saiu, 2017). Lastly, academia has been doubting the very existence of the smart city, questioning if smart urbanism is mere *corporate storytelling* or a tangible urban phenomenon, a utopia or an actually existing character of the city of the twenty-first century (Hollands, 2008; Shelton et al., 2015; Söderström et al., 2014).

This chapter argues that the key characteristics, ideas, tensions and contradictions of smart urbanism have deep historical and philosophical origins whose exploration can explain not only where the smart city comes from but also what urban futures this ideal is leading to. The aim of the following sections is to examine the theoretical underpinnings of the smart-city phenomenon, approaching it as part of a much broader and older phenomenon in the history of the city: *modernity*. First, the chapter explores Francis Bacon's *New Atlantis* as a seminal example of an experimental city in which technology is employed to understand, rationalize and control the built and the natural environment. Here the book unpacks the notion of modernity, emphasizing its impact on the conceptualization and development of the city. Second, the same conceptual thread is followed throughout the Second Industrial Revolution and the diffusion of *modernism*. In this part of the chapter, the narrative cuts across Nietzsche's philosophy and the visual arts of Expressionism and Futurism to shed light on the bipolar experience of the modernist city as a cradle for both order and chaos. Third, the chapter moves to Bauman's notion of *light modernity* to illuminate the diffusion of ICT and, above all, its profound but scarcely visible impact on the design and life of cities. At this juncture, the focus splits between the real-life cases of Los Angeles and Singapore as the pioneers in smart urbanism, and the abstract visions of digital cities cultivated by thinkers like Pascal and Mitchell. As the chapter concludes, the smart city lies in an ambiguous terrain between the real and the fictional, where castles in the air can be easily mistaken for down-to-earth urban spaces.

Modernity and the smart city

From a philosophical point of view, the origins of the smart city can be traced back to Francis Bacon's *New Atlantis*. Published in 1627, Bacon's utopian novel contains what are arguably the first images of an experimental city permeated by technology, where technological innovation is actively employed to produce information about the natural and the built environment. In the city described by Bacon, this information is, in turn, employed to rationalize and control the environment (intended as the totality of the surrounding conditions) for the development of human societies. The narrative of the book follows a classical *topos* in utopian literature: the journey to a mysterious land. A European ship is attempting to reach China and Japan from Peru, but the crew loses the route somewhere in the Pacific Ocean due to strong winds. 'So that finding ourselves, in the midst of the greatest wilderness of waters in the world, without victual, we gave ourselves for lost men and prepared for death' the narrator tells (Bruce, 1999: 152). However, at dawn, the sailors discover that the sea has brought them within sight of an uncharted island. Upon landing, they quickly learn that the island is a city-state called Bensalem which is ruled by a community of scientists working for Salomon's House, a large and powerful research centre.

Bensalem, under the aegis of Salomon's House, is a place of untamed and ubiquitous technological innovation. The city itself is a vast living laboratory where new technologies are being constantly implemented and tested. In the name of progress, a number of futuristic experiments are regularly conducted in the open air by the local community of scientists, and machines are part of everyday life. Among the creations of the House of Salomon there are submarines, flying devices and robots reminiscent of the mechanical artefacts invented by Leonardo da Vinci. It is important to emphasize that in Bacon's ideal city, science and technological innovation are seen as instruments whereby the human condition can be improved (Mumford, 1962). In this sense, technological development is not simply associated with urban development, but also with human development. This is a key intellectual passage because, for the first time, the city is clearly depicted and conceptualized as a rational space that humans control and craft through technology to improve their lives. To explore this point in more depth, an overview of Bacon's broader philosophical system is needed.

The ideas exposed in *New Atlantis* are largely based on Bacon's earlier work *The novum organum*, fully *Novum organum scientiarum (New instrument of science)*. In this text the English philosopher and scientist conceptualizes science as an instrument in the hands of men, through which nature can be controlled and subjugated to human needs. Bacon's main concern is understanding the causes behind natural phenomena such as solar radiation and heat. Like Aristotle, he believes that *verum scire est scire per causas* (to truly know is to know causes): a belief motivated by the desire to master the mechanics of nature as a way to reproduce or stop a particular natural phenomenon at will (Fowler, 1878). Back in the early seventieth century, these ideas were revolutionary with their portrayal of the *anthropos* as the forger of destiny: the destiny of the individual and that of the world itself. Initially cultivated in the Renaissance and later expanded in the Enlightenment, this philosophy of personal and global development is commonly called *modernity*. According to Burckhardt (1965), the modern anthropos arises at the end of the Middle Ages breaking the then-dominant view of life as a rigid experience shaped by uncontrollable forces, such as deities and nature. The condition of modernity is about dreaming of the possibility of an alternative reality, and realizing that dreams can become true by unleashing the power of technology to materialize them (Berman, 2000).

Prior to Bacon's work, evidence of modernity, as a ground-breaking system of thinking, philosophy of life and *modus operandi*, emerges in several Renaissance texts. In his *De hominis dignitate (Oration on the dignity of man)*, written in 1486, Pico della Mirandola (1996), for instance, describes the human being of his time as *faber fortunae suae*: the maker of his own destiny. Similarly, Machiavelli (1985) in his seminal work *The Prince*, originally printed in 1532, claims that fortune governs over only half of life, and the rest can be actively controlled by human free will. This modern attitude towards development is also evidenced in changes in urban development. While the planning and design of the Medieval city tended to accommodate the morphology of the natural environment, the Renaissance city is not afraid to alter the landscape, in order to accommodate the will of the urban developer (Mumford, 1961). This is the time when, particularly in Europe, the art of urban design begins to follow the practise of land levelling, turning the land into a blank canvas for urbanists to realize their dreams (Conforti, 2005; Mumford, ibid.). Moreover, as noted by Kruft (1989), in the Middle Ages there is little or no trace of plans for ideal cities. The visionary and experimental urban initiatives of societal change discussed in Chapter 1, are a trademark of the Renaissance and, as such, a product of modernity.

Seen from this perspective, *New Atlantis* can be understood as the first theory of experimental urbanism, capturing the force of modernity. In Bacon's ideal city, urban developers fully control the built and the natural environment through technology to foster progress. Dreams are believed to be within reach and urban experiments redefine the boundaries of reality. As in every piece of utopian literature, *New Atlantis* was never realized. However, since its publication, modernity has been constantly and deeply influencing the conceptualization and development of cities. Across time and space there are two bursts of blended technological and urban development, in particular, which are relevant to the history and philosophy of smart urbanism. The first one, the focus of the next section, originates in the Second Industrial Revolution and in the broad cultural phenomenon that has gone down in history as *modernism*. This is a period of revolutionary inventions in established as well as new fields of knowledge, characterized by an unprecedented scale of technological innovation in urban contexts.

Modernism and the smart city

There is a substantial difference between the First Industrial Revolution and the Second Industrial Revolution. The former had been driven mainly by amateurs (Arkwright, for instance, the inventor of the spinning frame was a barber; Cartwright, the mind behind the power loom, was a clergyman; Watt, before inventing the Watt steam engine, was working as an instrument maker). The latter instead saw a strong synergy among science, industry and economy, with teams of professional scientists working in partnership with private companies. During the years of the Second Industrial Revolution, from the late nineteenth century throughout the early twentieth century, scientific research and technological innovation were largely supported by capital investments. The logic of profit had just married the logic of mass-production: a union meant to maximize the return on investment. As Jensen (1993: 834) notes, the technological changes of this epoch were characterized by 'a shift to capital-intensive production' and 'rapid growth in productivity' (see also Chandler, 1990). These factors, coupled with novel devices of mass-distribution, led particularly in the United States and Europe to a pervasive diffusion of new technologies which became part of the everyday life of cities, thereby influencing their design and experience.

The production of steel, for example, an alloy of iron and carbon, which had been known since antiquity, became less expensive in 1856 with the invention of the Bessemer converter and, together with the introduction of reinforced concrete in civil engineering in 1884, it led to the construction of monolithic architectural structures such as skyscrapers and long suspension bridges. In 1885,

Benz, a German designer and engineer, patented the first automobile and in 1888 Dunlop, a Scottish inventor, produced early pneumatic tyres. A few decades later, these two inventions were combined and popularized by the Ford Motor Company (1903), with substantial repercussions on both the built and the natural environment. The flow of automobiles in urban settlements, as a new dominant form of transport, led to the development of highways and arterial roads which expanded the fabric of cities, taking up large amounts of space. Prime farmland and ecosystems such as forests, rivers and lakes were lost in the process and, from their ashes, hitherto unimaginable urban landscapes emerged (Kenworthy and Laube, 1996).

What Hall (2002: 295) terms 'the city on the highway' came with new forms of mobility which were quickly translated into new geographies of spatial organization. As Hall (2002) notes, car-owners became able to travel long distances to get access to urban services such as retail, education and health, as well as to their workplace. For them, living in or close to the city centre stopped being a necessity. Housing units were consistently built away from the heart of the city, through intense processes of suburbanization which expanded the built environment horizontally. Suburbs had already been present since another transport revolution, the train, but the car marked the beginning of a novel lifestyle based on private ownership, highly dependent on fossil fuels and responsible for much of the sustainability issues discussed in Chapter 1. At a smaller scale, the whole phenomenon was clearly reflected in architecture, particularly in the work of Frank Lloyd Wright who in 1909 (six years after the genesis of the Ford Motor Company) pioneered new models of domestic space with *Robie House*: an American suburban villa featuring a three-car garage.

The spirit of this epoch, the *zeitgeist*, can be understood more in-depth by examining the connections among the then mainstream philosophies, paintings and architectures. In Germany, for instance, art was being influenced by the philosophical system of Nietzsche (1844–1910) who 'overthrew the Apollonian or balanced image of Greek culture, in favour of a Dionysian image of intoxicating dynamism' (Watkin, 2005: 588). In *The birth of tragedy*, Nietzsche (2008) draws upon Greek mythology to illustrate the contrast and tension between two sides of human nature. One side is exemplified by Apollo, son of Zeus, and the god of rationality and order. The other is represented by Dionysius, who was also the son of Zeus, but he had a personality that was the opposite of his brother. This is why Dionysius became the patron of irrationality and of instinctive and emotional behaviours. The Dionysian side inspired Expressionism: the artistic movement championed by painters such as Wassily Kandinsky (1866–1944) and Paul Klee (1879–1940), in which the environment that is portrayed is distorted by the emotions of the author. Through Expressionist visual art everything is seen exclusively through a subjective perspective which is superimposed on the materiality of the world. Similarly, in Expressionist architecture the built environment is designed and shaped by the emotions and will of the architect whose personality is superimposed on the natural environment.

Typical of this strand of architecture, which was particularly popular in Germany in the 1920s, are large towering buildings whose construction was made possible by the development of new building materials and engineering solutions. The style of prominent Expressionist architects like Hans Poelzig and Bruno Taut manifests an individualistic take on design and the impulse to leave history behind in the attempt to look modern and different. In this sense, Expressionist architecture strongly echoes the *fin-de-siècle* philosophy of Nietzsche, which was explicitly mentioned by some Expressionist artists of the built environment. Taut, for instance, as noted by Pehnt (1973), was often quoting Nietzsche in his sketches, referring in particular to *Thus spoke Zarathustra*. In this seminal book, published in four volumes between 1883 and 1891, the German philosopher introduces the concept of the *Übermensch* (Superman): an individual who fights to express personal values by acting against conventionalism and towards self-actualization (Nietzsche, 2005). Nietzsche's ideal human being, like a modern Dionysius, is free and not afraid of the turbulent sea of emotions kept inside. As Nietzsche (2005: 15), through the voice of Zarathustra, claims 'one must still have chaos within oneself, to give birth to a dancing star.' In a similar vein, Expressionist architects were seeking to break the chains of the past in order to free themselves from history and its heritage, and fully express their individuality. In this cultural and philosophical context, architecture becomes the culmination of the ego intended as the self: an entity assumed to be different from the rest of the world and other selves, and indifferent to such difference.

In the context of the Second Industrial Revolution, the voices of Nietzsche and the Expressionists reveal a key element of the cultural trends and transformations of the turn of the century, broadly grouped under the term *modernism*. This element is *paradox*. On the one hand, as scholars such as Harvey (1989) and Habermas (2018) remark, modernism derives from the ideas of modernity discussed in the previous section. In this sense, modernism is about rationality, order, progress, human development, construction, science and technology. On the other hand, however, modernism is also chaos, instinct, destruction, individualism and Dionysian ecstasy. Berman (2000) is adamant in his claim that the experience of modernity is ultimately paradoxical because of its almost bipolar nature. Similarly, Harvey (1989) draws upon Baudelaire's recognition that one half of being modern is being ephemeral, chaotic and discontinuous to assert that modernism cannot be conceptually pinned down to sole rationality. This tension is particularly evident in Futurism, a strand of modernism established in Italy in 1909, celebrating the union of seemingly contrasting ideas, priorities and dimensions: order and chaos, control and frenzied velocity, and the city and the individual.

From an artistic point of view, the visual emblem of these tensions can be seen in the art of Umberto Boccioni, which glorifies the age of the machine established through the Second Industrial Revolution, with a strong sensitivity to the built environment. In *La città che sale* (*The City Rises*), painted between 1910 and 1911, Boccioni depicts the expansion of the city, the expansion of progress and the expansion of industry through a single multiform image (see Figure 3.1). This futurist painting offers three interconnected scenes. In the upper part, on the left

FIGURE 3.1 Study for The City Rises (1910).
Source: MoMA

side, an industrial landscape represents the historical context of Boccioni's work. Factories and power plants are pumping capital and energy into the city, thus sustaining its growth which is depicted on the right side. Here a construction site is animated by a number of workers, through whose intense labour the city is growing vertically, aiming for the sky, without limits. The core of the painting is occupied by an amalgam of horses and men which are melted together in an act of extreme dynamism. This vision symbolizes the contradictions of modernism. The horses are the raw power of industry and technology, unleashed to build the city. They are possessed by a supernatural fury. Some men are trying to tame them, but theirs is a futile effort. Many are falling and being trampled to death. Others have given up and are sucked into the vortex. In this maelstrom lies the paradox and tragedy of modernity. The city is a rational space, controlled and crafted by humans, where modern technologies are instruments of human development. Simultaneously, the city is a chaotic place where individuals jump into the source of the power of technological innovation for individual gains. They do not know if they will be capable of controlling it, but they do not care, as they do not care about the havoc that their action is wrecking in the surrounding environment.

At a smaller scale, the feelings expressed in Futurism are immortalized in *La strada entra nella casa* (*The Street Enters the House*). In this painting, Boccioni uses intersecting planes to visualize multiple perspectives and bring different scenes to life, on the same canvas (see Figure 3.2). The result is a Dionysian visual ecstasy picturing a modern bacchanalia. In the central scene, a group of workers, surrounded by piles of red

FIGURE 3.2 The Street Enters the House (1911).
Source: Wikimedia Commons

bricks, are frenetically constructing a new building. The city is already hyper dense, and the sky can scarcely be seen. Standing on her balcony is a woman. She does not close the windows that separate her home and her life from the hectic life of the city. She leaves them wide open to let the noise and chaos of urban living in, like she was attending an enthralling concert performance. Urban development's order and rationality are fused with an irregular and fragmented built environment which seems to be animated by a life of its own. Sonically, this is the same spirit that, in its most extreme form, pervaded the work of futurist musicians such as Luigi Russuolo. In *L'arte dei rumori* (*The Art of Noises*), a manifesto dated 1913, the Italian composer theorized the evolution of the human ear in relation to the evolution of the urban landscape. For Russuolo, the noise of the modern city had expanded the spectrum of sounds existing in nature, and was an exciting new music which required new instruments and compositions. Symbolic of this musical aesthetics is

Risveglio di una città (*Awakening of a City*), an experimental piece played with self-constructed noise-generating devices, which captures the ear as the sonic counterpart of the canvas painted by Boccioni.

It is also important to emphasize that the Dionysian cult of modernism was not esoteric and intended for small groups. It was celebrated and taught in and via the city through, for example, world's fairs which attracted millions of visitors. The Exposition Universelle of 1889 held in Paris with its *Galerie des machines* (*Gallery of Machines*), and Chicago's 1933 Century of Progress world's fair with its motto 'Science Finds, Industry Applies, Man Adapts,' exemplify the temples where the modernist spirit was being worshipped. Modernism was an exoteric doctrine and a popular lifestyle, with rites including new ecstatic forms of dancing such as the Charleston which was danced to high-tempo jazz music whose BPM (beats per minute) was, on average, twice that of ballet. Of course, as Berman (2000) stresses, the experience of modernity was not universally shared and embraced. In music, for instance, the diffusion of jazz was accompanied by the canonization of what is now called *classical music*. However, particularly from a technological and urban perspective, due to the synergy among science, industry and private companies discussed earlier in this section, the turn of the century represents a global point of no return. In 1879, Edison invented the incandescent light bulb. This small artefact was followed by the so-called *War of the Currents* which introduced electricity into the built environment through new urban infrastructures (Klein, 2010). Cities began to be wired and their metabolism changed. This new infrastructural landscape powered the domestic consumption of several innovative devices like the telephone, the radio and the television which were soon mass-produced. Such devices are early examples of ICT and the seeds of the 1970s digital revolution which was about to generate the technologies that are the basis of today's smart-city initiatives.

Light modernity and the invisible birth of the smart city

The Second Industrial Revolution had accumulated significant momentum, and information and communication technologies developed rapidly throughout the first half of the twentieth century. In the 1970s, with the invention of computers, ICT reached a tipping point that was going to have a profound, but not immediately visible, effect on cities. This is the second burst of modernity in the conceptualization and development of the city, crucial for the understanding of smart urbanism. In his comprehensive analysis of the network society, Castells (2011) argues that the then convergence of micro-electronics, computing, telecommunications and broadcasting was paradigmatic because of the high degree of pervasiveness of the technological changes in question. By the time ICT devices such as computers were developed, the world had been substantially reshaped by two industrial revolutions, and large segments of the population had already been exposed to the daily use of technology and the value of information (Castells, ibid.). Therefore, while in the Renaissance the invention of the printing press, for example, was beneficial only to those few people who could read, the products of the ICT revolution managed to penetrate most if not all domains of human activity. This does not mean that the process was geographically and socially homogenous. As pointed out by Castells (2011: 33) himself, there were and still are several 'switched-off areas' where people do not have access to the power of ICT. However, looking at the big picture, Castells (2011: 6) sees a novel and widespread 'network architecture which, as its inventors wanted, cannot be controlled from any centre, and is made up of thousands of autonomous computer networks that have innumerable ways to link up, going around electronic barriers.'

Despite its ground-breaking character, Castell's network society did not shock the physical structure of cities. Although, just as happened in the Second Industrial Revolution, the technological discoveries of the early 1970s were quickly pushed forward by the capitalistic interests of competing private companies (Intel, IBM and Apple), the dynamics of urbanization remained largely untouched by ICT until the new millennium (Hall, 2002). This inertia in relation to urban design and urban planning, can be understood by unpacking what Bauman (2013) calls the passage from *heavy modernity* to *light modernity*. For Bauman (ibid.), the concept of heavy modernity represents the age of a type of industry characterized by an obsession for size and space (both intended from a physical perspective). This is the time of the Second Industrial Revolution. A time dominated by large machines and large factories tied to the ground by means of steel and concrete: the 'era of territorial conquest' when 'wealth and power were firmly rooted or deposited deep inside the land – bulky, ponderous and immovable like the beds of iron core and deposits of coal' (Bauman, 2013: 114). Light modernity, on the other hand, symbolizes an epoch when, particularly from a technological point of view, 'space no more sets limits to action and its effects' (Bauman, 2013: 117). In this phase of modernity, the power of technology tends to extend far beyond the physical space where technology is installed. It travels at the speed of light, covering and cutting across such a wide range of spaces that it almost becomes ubiquitous.

In more practical terms, as noted in the previous section, the new technologies and materials of the Second Industrial Revolution, such as engines, cars, steel and reinforced concrete, were characterized by a marked weight, volume and, in essence, materiality. Their physical substance, therefore, had an immediate impact on cities, simply because it could not have been ignored. As in the seventeenth century the diffusion of carriages had led to the straightening of roads and the creation of sidewalks to protect pedestrians, so in the early twentieth century the morphology of cities had to adapt to the presence of cars and to their materiality (Mumford, 1961). Such changes were not necessary in the aftermath of the ICT revolution whose main technological breakthroughs were far less tangible. The Internet and its flows of data have an ethereal essence, while computers, since their genesis, have been evolving through processes of miniaturization in order to occupy less and less physical space. From a physical point of view, the touch of light modernity was much softer than that of heavy modernity and, in comparison, the city barely felt it.

The resistance of the city to the changes brought by the ICT revolution can also be observed from an architectural point of view. In the 1970s, while computers were being developed and popularized, the dominant movement in architecture was the so-called *high-tech*. Promoted in particular by three starchitects, Norman Foster, Renzo Piano and Richard Rogers, high-tech architecture was characterized by heavy and large steel-based structures which were philosophically very far from the idea of light modernity expressed by Bauman. The first product of this architectural style was the iconic Pompidou Centre (1971–1977) in Paris, designed by Piano and Rogers. The building reveals the machine that is behind the architecture, with a façade made of functional pipes and cables which are coloured, emphasized and exposed. It must be remembered that fundamental to the implementation of the project was the work of Peter Rice who was not an architect, but a structural engineer. This was a type of architecture that was not supposed to marry the newborn science of computers. Because of its love for imposing buildings, like Foster's HSBC headquarters in Hong Kong (1979–1986), it had already married engineering. Such architecture was not spaceless and ethereal like the invisible flows of computer data. It was heavily bounded to a specific space and thus needed complex structures to sustain itself. In antithesis with the novel and rapidly expanding technologies of the information age, high-tech, as Glancey (2003: 205) observes, displayed nostalgia for the modernist architecture of the turn of the century, and ultimately it 'celebrated technologies that were passing rapidly into the history books.'

FIGURE 3.3 The façade of the Pompidou Centre. Heavy machinery on display, impossible to go unnoticed.
Source: Sebastien Devigne

However, as history is the sum of nonlinear processes, there are exceptions to take into account, and these are the cities that pioneered smart urbanism. Historically, the first exception is arguably Los Angeles which in the 1970s was a pioneer in what is now called *big data*. A policy report discovered by Vallianatos (2015), demonstrates that in 1974 the urban development of Los Angeles was being shaped by a large volume of computer data. Back then, a special administrative division, the Community Analysis Bureau, employed state-of-the-art computer technologies to store and organize information on different aspects of the city, such as traffic, housing, crime and poverty. The information was clustered in various data sets through the support of computers and eventually mapped out. The aim was to provide local policy-makers with a broad-based knowledge of the conditions of Los Angeles for them to make an informed decision on its development. Not long after LA's experimental urbanism, in the late 1980s, Singapore pioneered the concept of the *wired city*. Batty (2012: 191) recalls that 'the notion of wiring the city then was essentially one of providing networking for very diverse activities without any very specific uses in mind.' In practise, from a physical perspective, this involved the installation of fibre optic cables across the city, to create a network which, by the mid-1990s, was already producing and circulating data. The ICT infrastructure of Singapore was being used, for example, to decentralize business activities by allowing people to work from home, to implement a system of automatic payment on the road via scanners and smart cards, and to maximize the communication between citizens and the government through online portals (Arun and Teng Yap, 2000).

Singapore's experiment proved that marrying ICT and urban development was both possible and profitable. A few ICT companies quickly received and circulated the message using the term *smart city* explicitly. As pointed out by Vanolo (2014, 2017), Cisco, for instance, adopted the smart-city brand in the late 1990s when it began to offer its services to cities seeking to integrate ICT into their infrastructures. The moniker *smart city* was subsequently filed by IBM in 2009 and then registered as the company's trademark in 2011 (Anthopoulos, 2017; Söderström et al., 2014). Since the beginning of the twenty-first century, smart-city discourses have been proliferating, featuring in a number of local, national and supranational urban initiatives. At the same time, this is also when the smart-city phenomenon started to become ambiguous, elusive and hard to capture conceptually and empirically. A great number of smart-city initiatives came with a great deal of confusion over what these initiatives were exactly referring to, beyond a generic marriage between ICT development and urban development. One reason for such confusion is that the phenomenon of the smart city has taken place across different planes of reality and fiction, which have often been confused with each other. These planes manifest resemblance; they might also promise the same thing, but are ultimately different, and this difference matters as it draws a line between *existence* and *void*, and *life* and *death*.

When, at the turn of the millennium, smart-city initiatives begun to appear around the world, ideas of an urbanism centred around ICT were being cultivated by several thinkers on cities and technologies in abstract and, at times, speculative terms (Angelidou, 2015; Cugurullo, 2018b). In this regard, Graham (2004) highlights the contribution of media theorist Marshall McLuhan who in 1964 predicted the dissolution of the traditional city due to the pressure exerted by an increasingly interconnected world. In a similar vein, Pascal (1987) advanced the term *vanishing city* which is akin to Pawley's (1995) notion of *spaceless cities*. These abstract ideas, as Marvin (1997) notes, promoted the dematerialization of traditional urban societies in favour of decentralized digital societies where socio-economic relations occur remotely and are enabled by information and communication technologies. Here, it is crucial to recognize that grammar can often be misleading. Although, in this type of literature, the present tense is frequently used by the authors, texts rarely engage with an ongoing action. Instead, the text tends to express a possibility. In other words, this is a literature which theorizes *the possible*, rather than describing *the actually existing*.

Within this genre of urban scholarship, the most prominent thinker was arguably William J. Mitchell who expressed an urbanism in which the built environment merges with the digital environment. For Mitchell (1995, 1999), ICT creates virtual spaces characterized by new geometries and temporalities which are not in sync with traditional urban forms. As virtual spaces become more prominent in societies, cities, he argued, must adapt in order to accommodate these new digital spatialities. Visionary in nature, Mitchell's (1999: 4, 8) urbanism seeks to 'reinvent public places, towns and cities' and 'rethink the role of architecture.' His *e-topias*, defined as 'electronically serviced, globally linked cities,' are equally built with concrete and chips forming a mix of physical and digital environments (Mitchell, 1999: 8). In these cities, people do not have to leave their home to work or meet someone, because physical streets have been substituted by virtual highways. Moreover, key resources such as energy and water are managed automatically by the software that underpins the city, which reacts according to data produced in real-time by sensors embedded in the built environment. Waste is avoided by what Mitchell (1999) calls *intelligent operations* and *smart systems*, and so are human mistakes thanks to the superior calculation skills of computers. However, as convincing, plausible and realistic as Mitchell's vision might sound, at the end of the day, this is a mental image produced by an imagination which does not automatically mirror reality.

Nowadays, the trend has not changed. On the one hand, as highlighted in the first section, the smart city is a global model of city-making which is being adopted by an increasing number of cities. As such, the smart city has an actual dimension and a verifiable existence. On the other hand, however, ideas of smart urbanism are simultaneously being developed in a theoretical manner by contemporary urbanists, architects and philosophers. In *The city of tomorrow*, for example, Ratti and Claudel (2016) envision urban spaces whose energy, transport and security systems are optimized and automated by smart technologies. As the title of the book hints, the spaces that are being discussed are not spaces of today, but rather of a future whose exact unfolding cannot be predicted. This does not mean of course that visions of smart cities are always impractical or not grounded in science and empirical research. Yet, they are often hypothetical and, therefore, do not necessarily have actually existing manifestations. At this junction in time, examples of theoretical smart cities abound. Umbrellium (2019), for instance, a London-based private company, has developed a smart pedestrian crossing on a responsive road surface, which appears only when pedestrians are in its proximity. This smart urban technology exists in the mind of the developers, as well as in the paper and websites where it is portrayed. It even exists temporarily in the real urban spaces where it is being tested, and tomorrow it might become a permanent feature of the infrastructure of a city. Today it is not, along with many other instances of smart urbanism.

Conclusions

The theoretical underpinnings of smart urbanism are considerably older than what present smart-city initiatives superficially suggest. The smart-city phenomenon is composed of images of modern, high-tech and futuristic cities which seek to leave history behind in the attempt to look novel. However, this chapter has shown that smart urbanism is intrinsically linked to a past whose heritage is still dominant. History has not been left behind. Conceptually, the notion of *smart cities* resonates with Bacon's *New Atlantis* and early ideas of modernity which picture technology as an instrument of progress, capable of controlling the built and the natural environment. The thread then thickens with the advent of modernism and the widespread technological innovations triggered by the Second Industrial Revolution which consecrated the progressive role of technology in the city. Subsequently, technology and its urban manifestations have changed, particularly with the diffusion of ICT, but the logic has remained largely untouched. Like its historical predecessors, the smart city continues to profess and practise a universal credo of modernization.

However, while today the smart-city phenomenon, with its escalation of modernization propelled by the economic interests of ICT companies, can easily leave the technologies of the past behind, it cannot cut off and leave behind its theoretical lineage. This is because modernization has been technological and not philosophical. The dominant ideas of modernity and modernism underpinning the genesis and development of smart urbanism have not been sufficiently challenged to the point of retreating or morphing. As a result, the tensions and contradictions that smart-city agendas now manifest are not new: they are the recurring spectres of an ever-lasting past. Inside present smart-city projects pulse the bipolar personality of Nietzsche's modern *anthropos* seeking rationality while embracing chaos, the individual interests of private companies hidden under the mask of global development, and the furious horses depicted by Boccioni bringing chaos under the name of order. They are all there.

Above all, like *New Atlantis*, the smart-city phenomenon is animated by the spirit and curse of utopia. In this sense, smart urbanism has a vein of *utopianism* which, as Sargent (1994) explains, is a form of *social dreaming*. Visions of smart cities are visions of cities perfected by technology, sustainable societies whose endless needs are fulfilled by endless technological innovation. Still, dreams can be depicted as universal but, in the end, they are a personal experience, and whose needs smart-city initiatives are realizing and why are questions that necessitate urgent empirical research. Furthermore, regardless of who owns and exploits them, dreams might never come true. The smart city lies in a dangerous terrain between the real and the fictional, the project and the theory, the visible and the invisible. When these oppositional dimensions collide, what emerges is a complex semi-utopian city equally made of infrastructures and spaces that are, never were, might be and never will. Ultimately, smart urbanism can be a potent driver of change, pushing cities towards the realization of an ideal, while also being a mirage. An illusion crafted and kept far away on purpose. It is time to reach for and touch alleged smart cities and, if necessary, to wake up.

References

Albino, V., Berardi, U. and Dangelico, R. M. (2015). Smart cities: Definitions, dimensions, performance, and initiatives. *Journal of urban technology*, 22 (1), 3–21.

Allwinkle, S. and Cruickshank, P. (2011). Creating smart-er cities: An overview. *Journal of urban technology*, 18 (2), 1–16.

Angelidou, M. (2015). Smart cities: A conjuncture of four forces. *Cities*, 47, 95–106.

Angelidou, M., Psaltoglou, A., Komninos, N., Kakderi, C., Tsarchopoulos, P. and Panori, A. (2018). Enhancing sustainable urban development through smart city applications. *Journal of Science and Technology Policy Management*, 9 (2), 146–169.

Anthopoulos, L. G. (2017). *Understanding Smart Cities: A Tool for Smart Government Or an Industrial Trick?*. Springer, New York.

Arun, M. and Teng Yap, M. (2000). Singapore: the development of an intelligent island and social dividends of information technology. *Urban Studies*, 37 (10), 1749–1756.

Barns, S., Cosgrave, E., Acuto, M. and Mcneill, D. (2017). Digital infrastructures and urban governance. *Urban Policy and Research*, 35 (1), 20–31.

Batty, M. (2012). Smart cities, big data. *Environment and Planning B: Planning and Design*, 39, 191–193.

Batty, M. (2013). Big data, smart cities and city planning. *Dialogues in Human Geography*, 3 (3), 274–279.

Batty, M. (2016). Big data and the city. *Built Environment*, 42 (3), 321–337.

Bibri, S. E. and Krogstie, J. (2017). Smart sustainable cities of the future: An extensive interdisciplinary literature review. *Sustainable Cities and Society*, 31, 183–212.

Bauman, Z. (2013). *Liquid modernity*. John Wiley & Sons, Oxford.

Berman, M. (2000). *All that is solid melts into air: the experience of modernity*. Verso, London.

Bina, O., Inch, A. and Pereira, L. (2020). Beyond techno-utopia and its discontents: On the role of utopianism and speculative fiction in shaping alternatives to the smart city imaginary. *Futures*, 115, 102475.

Bruce, S. (Ed.). (1999). *Three early modern Utopias. Utopia, New Atlantis and The house of pines*. Oxford. University Press, Oxford.

Burckhardt, J. (1965). *The civilization of the Renaissance in Italy*. Phaidon, London.

Burns, R., Fast, V., Levenda, A., and Miller, B. (2021). Smart cities: Between worlding and provincialising. *Urban Studies*. doi:10.1177/0042098020975982.

Caprotti, F. (2019). Spaces of visibility in the smart city: Flagship urban spaces and the smart urban imaginary. *Urban Studies*, 56 (12), 2465–2479.

Caragliu, A., Del Bo, C. and Nijkamp, P. (2011). Smart cities in Europe. *Journal of urban technology*, 18 (2), 65–82.

Cardullo, P. and Kitchin, R. (2019). Smart urbanism and smart citizenship: The neoliberal logic of 'citizen-focused' smart cities in Europe. *Environment and Planning C: Politics and Space*, 37 (5), 813–830.

Carvalho, L. (2014). Smart cities from scratch? A socio-technical perspective. *Cambridge Journal of Regions, Economy and Society*, 8 (1), 43–60.

Castells, M. (2011). *The rise of the network society: The information age: Economy, society, and culture* (Vol. 1). John Wiley & Sons, Oxford.

Chandler, A. D. (1990). *Scale and scope: The dynamics of industrial capitalism*. Harvard University Press, Cambridge.

Colding, J. and Barthel, S. (2017). An urban ecology critique on the 'smart city' model. *Journal of Cleaner Production*, 164, 95–101.

Coletta, C., Evans, L., Heaphy, L. and Kitchin, R. (Eds.). (2018). *Creating Smart Cities*. Routledge, London.

Conforti, C. (2005). *La citta' del tardo Rinascimento*. Laterza, Roma.

Cowley, R., Joss, S. and Dayot, Y. (2018). The smart city and its publics: insights from across six UK cities. *Urban Research & Practice*, 11 (1), 53–77.

Cugurullo, F. (2018a). Exposing smart cities and eco-cities: Frankenstein urbanism and the sustainability challenges of the experimental city. *Environment and Planning A: Economy and Space*, 50 (1), 73–92.

Cugurullo F. (2018b) The origin of the Smart City imaginary: from the dawn of modernity to the eclipse of reason. In Lindner C. and Meissner M. (Eds.) *The Routledge Companion to Urban Imaginaries*. London, Routledge.

Cugurullo F (2020) Urban Artificial Intelligence: From Automation to Autonomy in the Smart City. *Front. Sustain. Cities* 2, 38. doi:10.3389/frsc.2020.00038.

Dameri, R. P., Benevolo, C., Veglianti, E. and Li, Y. (2019). Understanding smart cities as a glocal strategy: A comparison between Italy and China. *Technological Forecasting and Social Change*, 142, 26–41.

Della Mirandola, P. (1996). *Oration on the Dignity of Man*. Gateway Editions, Washington D.C.

Fowler, T. (Ed.). (1878). *Bacon's Novum Organum*. Clarendon Press, Oxford.

Glancey, J. (2003). *The Story of Architecture*. Dorling Kindersley, London.

Glasmeier, A. and Christopherson, S. (2015). Thinking about smart cities. *Cambridge Journal of Regions, Economy and Society*, 8, 3–12.

Glasmeier, A. and Nebiolo, M. (2016). Thinking about smart cities: The travels of a policy idea that promises a great deal, but so far has delivered modest results. *Sustainability*, 8 (11), 1122.

Goodspeed, R. (2015). Smart cities: moving beyond urban cybernetics to tackle wicked problems. *Cambridge Journal of Regions, Economy and Society*, 8 (1), 79–92.

Graham, S. and Marvin, S. (2002). *Splintering urbanism: networked infrastructures, technological mobilities and the urban condition*. Routledge, London.

Graham, S. (Ed.). (2004). *The cybercities reader*. Routledge, London.

Grossi, G. and Pianezzi, D. (2017). Smart cities: Utopia or neoliberal ideology? *Cities*, 69, 79–85.

Haarstad, H. (2017). Constructing the sustainable city: Examining the role of sustainability in the 'smart city' discourse. *Journal of Environmental Policy & Planning*, 19 (4), 423–437.

Habermas, J. (2018). *The philosophical discourse of modernity: Twelve lectures*. Polity Press, Cambridge.

Hall, P. (2002). *Cities of tomorrow*. Blackwell Publishers, Oxford.

Harvey, D. (1989). *The condition of postmodernity*. Blackwell, Oxford.

Höjer, M. and Wangel, J. (2015). Smart sustainable cities: definition and challenges. In Hilty, L. M. and Aebischer, B. (Eds.). *ICT innovations for sustainability* (Vol. 310). Springer, Berlin.

Hollands, R. G. (2008). Will the real smart city please stand up? Intelligent, progressive or entrepreneurial?. *City*, 12 (3), 303–320.

Imrie, R. and Lees, L. (Eds.). (2014). *Sustainable London?: The future of a global city*. Policy Press, Bristol.

Jensen, M. C. (1993). The modern industrial revolution, exit, and the failure of internal control systems. *The Journal of Finance*, 48 (3), 831–880.

Joss, S., Sengers, F., Schraven, D., Caprotti, F. and Dayot, Y. (2019). The smart city as global discourse: Storylines and critical junctures across 27 cities. *Journal of Urban Technology*, 26 (1), 3–34.

Kaika, M. (2017). 'Don't call me resilient again!': The New Urban Agenda as immunology… or… what happens when communities refuse to be vaccinated with 'smart cities' and indicators. *Environment and Urbanization*, 29 (1), 89–102.

Karvonen, A., Cugurullo, F. and Caprotti, F. (Eds.). (2018a). *Inside smart cities: Place, politics and urban innovation*. Routledge, London.

Karvonen, A., Cugurullo, F. and Caprotti, F. (2018b). Introduction: situating smart cities. In Karvonen, A., Cugurullo, F. and Caprotti, F. (Eds.). (2018). *Inside smart cities: Place, politics and urban innovation*. Routledge, London.

Kenworthy, J. R. and Laube, F. B. (1996). Automobile dependence in cities: an international comparison of urban transport and land use patterns with implications for sustainability. *Environmental impact assessment review*, 16 (4), 279–308.

Kitchin, R. (2014). The real-time city? Big data and smart urbanism. *GeoJournal*, 79 (1), 1–14.

Kitchin, R. (2019). The timescape of smart cities. *Annals of the American Association of Geographers*, 109 (3), 775–790.

Kitchin, R., Coletta, C., Evans, L. and Heaphy, L. (2018). Creating smart cities. In Coletta, C., Evans, L., Heaphy, L. and Kitchin, R. (Eds.). (2018). *Creating Smart Cities*. Routledge, London.

Kitchin, R. and McArdle, G. (2016). What makes Big Data, Big Data? Exploring the ontological characteristics of 26 datasets. *Big Data & Society*, 3 (1), doi:2053951716631130.

Klein, M. (2010). *The power makers: steam, electricity, and the men who invented modern America*. Bloomsbury, London.

Kong, L. and Woods, O. (2018). The ideological alignment of smart urbanism in Singapore: Critical reflections on a political paradox. *Urban Studies*, 55 (4), 679–701.

Kruft, H. W. (1989). *Städte in Utopia: die Idealstadt vom 15. bis zum 18. Jahrhundert zwischen Staatsutopie und Wirklichkeit*. CH Beck, Munich.

Lee, J. Y., Woods, O. and Kong, L. (2020). Towards more inclusive smart cities: Reconciling the divergent realities of data and discourse at the margins. *Geography Compass*, e12504. doi:10.1111/gec3.12504.

Leszczynski, A. (2016). Speculative futures: Cities, data, and governance beyond smart urbanism. *Environment and Planning A: Economy and Space*, 48 (9), 1691–1708.

Luque-Ayala, A. and Marvin, S. (2015). Developing a critical understanding of smart urbanism?. *Urban Studies*, 52 (12), 2105–2116.

McFarlane, C. and Söderström, O. (2017). On alternative smart cities: From a technology-intensive to a knowledge-intensive smart urbanism. *City*, 21 (3–4),312–328.

McNeill, D. (2015). Global firms and smart technologies: IBM and the reduction of cities. *Transactions of the Institute of British Geographers*, 40 (4), 562–574.

Machiavelli, N. (1985). *The Prince*. University of Chicago Press, Chicago.

March, H. and Ribera-Fumaz, R. (2016). Smart contradictions: The politics of making Barcelona a Self-sufficient city. *European Urban and Regional Studies*, 23 (4), 816–830.

Martin, C. J., Evans, J. and Karvonen, A. (2018). Smart and sustainable? Five tensions in the visions and practices of the smart-sustainable city in Europe and North America. *Technological Forecasting and Social Change*, 133, 269–278.

Marvin, S. (1997). Environmental flows: telecommunications and the dematerialisation of cities?. *Futures*, 29 (1), 47–65.

Marvin, S., Luque-Ayala, A. and McFarlane, C. (Eds.). (2015). *Smart urbanism: Utopian vision or false dawn?*. Routledge, London.

Mitchell, W. J. (1995). *City of bits: space, place, and the infobahn*. MIT Press, Cambridge.

Mitchell, W. J. (1999). *E-Topia: 'Urban life, Jim—but not as we know it.'* MIT Press, Cambridge.

Mora, L. and Deakin, M. (2019). *Untangling smart cities: From utopian dreams to innovation systems for a technology-enabled urban sustainability*. Elsevier, Amsterdam.

Mumford, L. (1962). *The story of Utopias: ideal commonwealths and social myths*. Viking, New York.

Mumford, L. (1961). *The city in history: Its origins, its transformations, and its prospects*. Harcourt, Brace & World, New York.

Nietzsche, F. (2005). *Thus spoke Zarathustra*. Oxford University Press, Oxford.

Nietzsche, F. (2008). *The birth of tragedy*. Oxford University Press, Oxford.

Pascal, A. (1987). The vanishing city. *Urban Studies*, 24 (6), 597–603.

Pawley, M. (1995). *Architecture, urbanism and the new media*. Mimeo, New York.

Pehnt, W. (1973). *Expressionist architecture*. Praeger, New York.

Raco, M. (2005). Sustainable development, Rolled-out neoliberalism and sustainable communities. *Antipode*, 37 (2), 324–347.

Ratti, C. and Claudel, M. (2016). *The city of tomorrow: Sensors, networks, hackers, and the future of urban life*. Yale University Press, New Haven and London.

Saiu, V. (2017). The three pitfalls of sustainable city: A conceptual framework for evaluating the theory-practice gap. *Sustainability*, 9 (12), 2311.

Sargent, L. T. (1994). The three faces of utopianism revisited. *Utopian Studies*, 5 (1), 1–37.

Silva, B. N., Khan, M. and Han, K. (2018). Towards sustainable smart cities: A review of trends, architectures, components, and open challenges in smart cities. *Sustainable Cities and Society*, 38, 697–713.

Shelton, T., Zook, M. and Wiig, A. (2015). The 'actually existing smart city.' *Cambridge Journal of Regions, Economy and Society*, 8 (1), 13–25.

Söderström, O., Paasche, T. and Klauser, F. (2014). Smart cities as corporate storytelling. *City*, 18 (3), 307–320.

Trencher, G. (2019). Towards the smart city 2.0: Empirical evidence of using smartness as a tool for tackling social challenges. *Technological Forecasting and Social Change*, 142, 117–128.

Umbrellium (2019). Starling crossing. [Online] Available: http://umbrellium.co.uk/initiatives/starling-crossing/ [Accessed 10 November 2020].

Vallianatos, M. (2015). Uncovering the early history of 'big data' and the 'smart city' in Los Angeles. Boom California. [Online] Available: https://boomcalifornia.com/2015/06/16/uncovering-the-early-history-of-big-data-and-the-smart-city-in-la/ [Accessed 10 November 2020].

Vanolo, A. (2014). Smartmentality: The smart city as disciplinary strategy. *Urban Studies*, 51 (5), 883–898.

Vanolo, A. (2017). *City branding: The ghostly politics of representation in globalising cities*. Routledge, London.

Watkin, D. (2005). *A history of western architecture*. Laurence King Publishing, London.

PART II
The experiment

'Winter, spring and summer passed away during my labours; but I did not watch the blossom or the expanding leaves, so deeply was I engrossed in my occupation.'

(Frankenstein, *Chapter 4*)

4

AN ECO-CITY EXPERIMENT

The case of Masdar City

Introduction

This chapter shifts the focus from the realm of ideas to the physical world. The development of projects for eco-cities is today a tangible phenomenon. Ideas of cities in balance with nature have been formulated for centuries. Images of built environments blending with the natural environment have been drawn and painted since the beginning of urbanization. Countless urbanists have recurrently fantasized about creating urban spaces inspired by the laws of ecology. In the twenty-first century, the eco-city appears to have become a reality. Across the world, new settlements are being built and existing cities are being regenerated under the *eco-city* banner (Bibri, 2020; Caprotti, 2014a; Cugurullo, 2018; De Jong et al., 2015; Joss et al., 2013; Lin and Kao, 2020; Rapoport, 2014; Wu, 2012). The developers of these alleged eco-cities claim to have discovered the formula for urban sustainability. Despite substantial geographical differences, eco-city initiatives advance universal models of sustainable urban development which, in theory, should solve the most pressing socio-environmental issues faced by the planet, while also being economically profitable. Climate change, resource scarcity and environmental pollution are not feared by eco-city developers, stakeholders and promoters. For them, the eco-city is the solution and it is simple and real.

In reality, however, the eco-city phenomenon is complex and ambiguous. On the one hand, eco-city initiatives are real. They exist and mobilize large quantities of capital, building materials and labour to create master-planned cities from scratch, to demolish existing infrastructures and to reshape entire districts. Examples abound. Masdar City (Abu Dhabi), Forest City (Malaysia), Songdo (South Korea), Rawabi (Palestine), Punggol (Singapore), K.A. CARE (Saudi Arabia), Tianjin and Chongming (China) are manifestations of the eco-city phenomenon and of its international character (Caprotti, 2014b; Chang et al., 2016; Chang and Sheppard, 2013;

Cugurullo, 2013a; De Jong et al., 2016; Shwayri, 2013; Xie et al., 2019). In addition, these initiatives are real inasmuch as they impact on both local and global socio-environmental systems. The implementation of an eco-city project, for instance, drains substantial national economic resources and relies upon international investments. Similarly, it necessitates the production and consumption of metals, plastics and, above all, energy coming from on-site and off-site sources. On the other hand, an eco-city project might not necessarily produce an ecological urbanism and achieve sustainability. Eco-city initiatives are urban experiments attempting alternative and untested urban solutions. Because of their experimental nature they tend to be risky endeavours, and an eco-city experiment supposed to protect regional eco-systems can eventually end up destroying them in unpredictable ways (Xie et al., 2020). The discourse of experimentation might also hide an agenda which is actually not about sustainable development at all. On these terms, even if an urban project is officially called an *eco-city*, there can be a stark discrepancy between the label and what that initiative is actually producing. This chapter aims to empirically examine an existing eco-city project in order to understand how, within this strand of experimental urbanism, sustainability is being practised and with what results.

The chapter employs Masdar City, an eco-city project under development in Abu Dhabi (United Arab Emirates), as a case study. The Emirati eco-city experiment is at the centre of the inquiry for three key reasons. First, the Masdar City project is one of the pioneers of the eco-city phenomenon. Officially launched in 2007, it is the first urban initiative developed and promoted under the banner of the eco-city. As such, it has opened the way for a number of eco-city projects heading towards a similar understanding and practise of urban sustainability. Exploring Masdar City, therefore, offers an insight into its numerous epigones. Second, the Masdarian formula for urban sustainability is circulating around the world, manifesting evident traits of what in urban studies and human geography is commonly referred to as *policy mobility* (McCann, 2011; McCann and Ward, 2011; Peck, 2011). The ideals of urbanism cultivated in Abu Dhabi are being actively exported to other countries by developers and stakeholders, thereby contributing to the formation of 'a fairly uniform and consistent set of ideas for enhancing the sustainability of urban development,' which is international in scale (Rapoport, 2015: 4; Rapoport and Hult, 2017). In light of the internationalization of the Masdarian model, its critical examination is an attempt to demonstrate, to urban planners, architects and policy-makers, what urban futures the blind application of this model is likely to produce. Third, Masdar City is an existing and mature eco-city project. Abu Dhabi's take on ecological urbanism is not simply a set of ideas: it is a practise which has been enduring for over ten years (Cugurullo, 2020; Griffiths and Sovacool, 2020). Eco-city theories have been partly materialized in the Emirati built environment. People are also living in Masdar City and, in essence, the project has been sufficiently developed to permit the empirical investigation of its genesis, implementation and outcomes.

The remainder of the chapter is organized as follows. The starting point of the exploration of Masdar City is its context. In the next section, the chapter examines the geography, politics and political economy of Abu Dhabi, discussing how the Emirati eco-city project is the by-product of a regional agenda of development meant to regenerate the local economy. In so doing, the narrative sheds light upon the economic imperatives that are the basis of Masdar City, pointing out that for the Abu Dhabi Government (and thus for the developers of Masdar City), sustainability is understood and practised largely in business terms. This aspect is explored in more depth in section three where the project's economic mechanics are exposed. Here the book shows how eco-city developers make money out of sustainability, by using the built environment as a laboratory for experimental clean technologies which are eventually commercialized and sold globally. Subsequently, the chapter reveals how the Masdarian business influences the planning process, dominating the development of the new city and disregarding what is not profitable. Sustainability becomes thereafter the focus of the critique. The major environmental and social tensions of the Masdar City project are discussed, and the limits of its business-centred urban model are emphasized. The journey to Masdar City and the chapter end with a reflection on the discrepancy between the theories of ecological urbanism illustrated in Chapter 2 and the practise of ecological urbanism in Abu Dhabi. In Masdar City, the prefix *eco* does not open the door to ecology, but rather to economics. Behind this door, the path that follows is too narrow to reach the socio-environmental complexity required by a sustainable urbanism. It is this narrowness that is the main lesson to be learned from the case of Masdar City. The project ultimately fails to achieve sustainability because the vision is too narrow. Like its historical antecedents, this is an urban experiment which advances a narrow vision of the good city incapable of looking beyond the subjective dream of its developers.

The context of Masdar City

The geographical context of the Masdar City project is Abu Dhabi, the largest and most powerful state of the United Arab Emirates (UAE). Located in the southeast of the Arabian Peninsula, the UAE is a federation of seven emirates. It was forged in 1971 by the then ruler of Abu Dhabi, Sheikh Zayed bin Sultan Al Nahyan, after almost two centuries of British hegemony. During his reign, Sheikh Zayed established Abu Dhabi as the political centre of the UAE, and capitalized on its vast oil reserves to trigger an unprecedented economic boom (Davidson, 2009; Morton, 2011). After Zayed's death in 2004, his son Sheikh Khalifa expanded the country's oil industry, placing Abu Dhabi among the top-ten oil producers in the world. In 2008, it was calculated that the export of oil was generating US$ 90 billion per year (Abu Dhabi Government, 2008). Part of this revenue was translated into major regional and overseas investments, by means of which Abu Dhabi managed to build a global portfolio of financial assets, including stakes in Manchester City Football Club, Barclays and Virgin Galactic for an estimated total of US$ 300 to 875 billion (Sovereign Wealth Fund Institute, 2013).

The fruits of this golden era were distributed among the local population. Abu Dhabi's policy-makers designed a strong welfare system to grant the nationals (approximately 15 per cent of the total population) luxurious standards of living which would have been unimaginable prior to the economic revolution brought about by Zayed. Before the discovery of petroleum in the 1960s, Emirati people used to ride camels, and live in humble tents in the desert. Later, when the oil era began, camels became just a matter of tourism, and Maserati and Ferrari cars started to rule freshly paved highways. Tents were quickly replaced by opulent gated communities surrounded by artificial green belts, and what was a small fishing village turned into a global metropolis of spectacular architecture (Cugurullo, 2013b; De Jong et al., 2019; Molotch and Ponzini, 2019; Ponzini, 2011; Ponzini et al., 2020). This transformation was rapid and drastic. According to Tatchell (2010), contemporary Abu Dhabi bears very little similarity with its pre-oil ancestor, in terms of lifestyle, wealth and aesthetics. Once sustained by fishing and pearl diving, the small Arabian settlement seems to have become an unstoppable economic powerhouse, launching 2.5 million barrels of oil into the market every day. Not even the global financial crisis of the late 2000s was apparently capable of hindering the growth of Abu Dhabi. When the credit crunch hit the world economy in 2008, the emirate's GDP per capita raised by 20 per cent, showing no sign of recession (International Monetary Fund, 2013; World Bank, 2013).

However, behind this apparent solidity, a number of interconnected development challenges threaten the present and future of Abu Dhabi. First, the economy of the emirate is largely based on oil which is a finite resource. While Abu Dhabi's government has declared to have enough oil for the rest of the century, there is ground to suppose that the royal family is overestimating its resources (Chapman, 2014). Moreover, even if the government's claim was correct and unproven oil reserves were found in the emirate, petroleum is ultimately destined to run out. For this reason, the almost monochromatic economy of Abu Dhabi cannot be kept in existence over an infinite period of time. Second, apart from oil, Abu Dhabi's natural resources are scarce. The regional shortage of fresh water, in particular, poses a difficult challenge. When, 50 years ago, Abu Dhabi was a small fishing village, the needs of the Emirati population were sustained by the local supply of groundwater. Twenty-first century Abu Dhabi, however, relies heavily on desalination: an energy-intensive practise which is in turn sustained by oil. Third, the total population of Abu Dhabi is rapidly growing and is expected to triple by 2030 due to injections of foreign workforce (Abu Dhabi Urban Planning Council, 2008, 2013). As a result, population growth is increasing the need for desalination which, consequently, is increasing the consumption of oil whose exhaustion appears to be only a matter of time. Fourth, although Abu Dhabi is unlikely to be directly affected by global climatic shifts, because of its geographical position it could face waves of migration from nearby countries, such as Bangladesh, which are directly affected by climate change in the shape of floods and cyclones (Luomi, 2009). These flows of climate refugees would then further impact on the regional scarcity of natural resources. In essence, the present Abu Dhabi does not have enough resources to sustain the future of Abu Dhabi.

Finally, it is important to emphasize the invisible precarity of the political structure of Abu Dhabi as a state. The emirate is governed by an authoritarian regime which, because of the absolute power of its ruler (the sheikh), falls under the category of *sultanism*. This is a political system in which everybody is subject to the unquestionable authority of the leader who plays and controls the game of politics in an undemocratic manner (Linz and Stepan, 1996; Weber, 1964). Sultanism is an expensive political system to sustain for the leader. In Abu Dhabi, UAE nationals live in a gilded cage made of generous economic benefits (Ali, 2010). They have few or no political rights, but the wealth that the government guarantees them is a strong incentive not to complain about their lack of freedom. This is a key reason why, during the storm of protests generated by the Arab Spring in the early 2010s, Abu Dhabi experienced not a single major public demonstration. Emirati citizens have little interest in going against the sheikh and the royal family as that would mean putting their wealth at risk. However, if the oil-based economy of Abu Dhabi collapsed, so would the welfare system at the basis of the Emirati gilded cage. If the latter cracked, Abu Dhabi might share the fate of countries like Libya, Syria, Egypt, Bahrain and Tunisia.

Aware of the above challenges, in 2008 Abu Dhabi's government launched a long-term development agenda called *Vision 2030*. This agenda envisions an ideal future for the emirate and sets up a plan to realize it. Vision 2030 is the vision of the sheikh and the royal family, picturing the development of Abu Dhabi according to the dreams of the ruling class. It is a potent exercise in imagination, with concrete economic, social and environmental consequences (Cugurullo, 2016a). Vision 2020 is divided into two interconnected parts laid out in two separate documents. The first part, *Economic Visions 2030*, engages with the economy of Abu Dhabi, while the second part, *Urban Planning Vision 2030*, focuses on the urbanization of the region, defining Abu Dhabi's key planning strategies and priorities. Both parts include the creation of new urban settlements such as Masdar City which, as the remainder of the chapter will show, plays a crucial role in the implementation of Vision 2030.

Economic Vision 2030 is the heart of Vision 2030 and fully encapsulates the philosophy of development cultivated by Abu Dhabi's government. The document presents a policy roadmap produced by a multinational taskforce, 'to guide the evolution of the economy of Abu Dhabi through to the year 2030' and ultimately 'build a sustainable economy' (Abu Dhabi Government, 2008: 1, 17). The concept of sustainability is a core part of the document. More specifically, the adjective *sustainable* is used extensively to describe the type of economic development envisioned by Abu Dhabi's policy-makers. In Economic Visions 2030, *sustainable* and *profitable* are used interchangeably, and *sustainability* and *economic profitability* become synonyms. In the document, the term *sustainable* always precedes the word *economy*, and refers to policy strategies capable of generating profit perpetually. According to the narrative of Economic Vision 2030, a sustainable development is a form of economic development which can be sustained or, in other words, maintained and kept in existence indefinitely. Since Abu Dhabi's economy is founded on petroleum, Economic Vision 2030 states explicitly that the current economic architecture of the emirate is unsustainable. The document advocates immediate change, invoking an economy in which the production of oil is not forgotten, but rather accompanied by alternative business activities. On these terms, sustainability is portrayed as a diversified economy made of heterogeneous sectors acting in concert, to maintain the country's high GDP per capita, without questioning the local lifestyle.

Building upon the above picture of sustainability, Economic Vision 2030 sets as a key policy target the development of additional non-oil strands of the economy, meant to facilitate the emirate's transition to a post-petroleum era. In practical terms, this target is translated into major investments in Research and Development (R&D) with a focus on clean technology, and into securing strategic partnerships with multinational cleantech companies (such as Siemens and General Electric). The expected outcome of this strategy is twofold. First, the Abu Dhabi Government seeks to develop and establish state-of-the-art sources of clean energy (solar power plants and geothermal stations, for instance). Second, investments in R&D, in addition to targeting energy security, are expected to diversify the local economy. The economic logic is linear: clean technologies are commodities that can be sold globally, and their commercialization constitutes for Abu Dhabi a new economic sector which expands and diversifies the Emirati economy.

The same expectation and understanding of sustainability are reflected in the second part of Vision 2030: Urban Planning Vision 2030. In this document, Abu Dhabi's government recognizes a deep interdependence between economic development and urban development. According to the Abu Dhabi Urban Planning Council (2008: 14), 'sustainable economic growth requires co-ordinated economic and planning strategies.' This line of thought is implemented in policy terms, through an ambitious programme of urbanization, targeting the construction of new cities and districts supposed to fulfil the targets set in Economic Vision 2030. Sustainability is again included as a central theme and narrowly interpreted to fit an economic rationale. The objective of Urban Planning Vision 2030 is the same objective of Economic Vision 2030, namely the formation of a diversified economy which can be kept in existence in the foreseeable future. The built environment is the medium through which this objective is pursued. Urban sustainability is conceptualized with a stress on the *economic*, and little or no emphasis is given to the *environmental* and the *social*. It is within this geographical and conceptual context that the Masdar City project emerges.

The business of Masdar City

Masdar City is a master-planned settlement under development from scratch in the desert of Abu Dhabi. It is a state-funded US$ 20 billion-project managed by the *Masdar Initiative*, a public company serving the Abu Dhabi Government as the developer of the new Emirati experimental city which is *de facto* property of the sheikh. Officially promoted as 'the world's most sustainable eco-city,' Masdar City represents, according to the Masdar Initiative (2018: no page), an urban model capable of achieving sustainability wherever it is applied, with no geographical limit. 'The model of Masdar City can be replicated anywhere in the world' argued one Masdar Initiative manager in an interview. 'By the time Masdar City will be finished, we will have enough experience to create similar cities all around the world. I am sure that governments in other countries will be very interested' he contended optimistically, stating that not only the Emirati eco-city project is 'the first city to be fully sustainable': it is also 'bringing cash.'

The construction of Masdar City began in 2008, following the launch of Vision 2030. However, its vision goes back to 2007 when the London-based architecture firm Foster + Partners (F+P), after winning an international competition, was commissioned by the sheikh to realize the master plan for a new eco-city. The project is currently expected to be completed by 2030. By then, Masdar City is supposed to occupy a surface of 6 km^2 and accommodate 50,000 permanent residents. Because of its small dimensions and position adjacent to existing urban areas, Masdar City can be understood as a district of Abu Dhabi, rather than an independent city as the name suggests (Cugurullo and Ponzini, 2018). Yet, compared with other Emirati urban ventures, Abu Dhabi's project for a new eco-city possesses such distinct mechanics, planning strategies and urban design features, that Masdar City stands out in its local context. These two sides, the uniqueness of the experiment and the traditionality of the context, are not necessarily oppositional. It is common for experimental urban projects to have components that are unique to the experiment, while also manifesting characteristics which reflect their setting. The remainder of the chapter explores the Masdar City project, paying attention to both its peculiarities and contextual influences.

The emirate's alleged eco-city is a tool designed and employed by the Abu Dhabi Government to realize Vision 2030. Its history, ideas, implementation and targets align with Economic Vision and Urban Planning Vision 2030. According to the developers, Masdar City will 'grow the non-oil's share of the emirate's economy,' an objective pursued through three key mechanisms (Masdar Initiative, 2012: 10). First, Masdar City is used as a living laboratory where private companies can research and develop new technologies. The process is linear and begins outside Masdar City. A cleantech company like Siemens, for example, has an idea for what it believes can be an innovative product. The German company decides to start a partnership with the Masdar Initiative, and moves to Masdar City where the built environment is employed as a lab. The deal is simple: the new city can be used to create and test the novel technology that Siemens has in mind, in a real-life environment. Once a partnership is forged, the Masdar Initiative shares its team of engineers and computer scientists, to assist in the development of the product. Moreover, the city itself offers all the infrastructures and spaces necessary to research, examine, manufacture and test state-of-the-art cleantech products under controlled conditions. Electron microscopy equipment, 3D printing facilities, advanced electrochemistry instruments, as well as offices, are all part of the Masdarian package which is strategically crafted by the Masdar Initiative to accommodate its partners' needs.

The partners of the Masdar Initiative are numerous. They include cleantech multinationals, such as Schneider, Mitsubishi, Siemens and General Electric, which have moved a portion of their business to Abu Dhabi's new city to develop a wide range of products ranging from smart grids to photovoltaics, and from automated transport systems to low-carbon building materials. A representative from Schneider stated in an interview: 'We have a strategic alliance with the Masdar Initiative. We test new technologies, smart grids, for example, and we develop them together.' As he further explained, 'the deal is that once a product is developed, it gets implemented into the city immediately.' As soon as a new technology becomes part of the urban fabric of Masdar City, its performance begins to be monitored and evaluated. The companies joining the Masdarian venture sell technologies which are ultimately meant to function in a city. Therefore, for them, the possibility of testing their products in the environment for which they have been designed, rather than in an indoor lab, is a unique opportunity. 'We want to test our products in a real-life environment,' answered a manager from Siemens when he was asked about his company's main reason for joining the Emirati eco-city project. The flow of data on how the experimental technologies tested in Masdar City perform is incessant. Data collection is automated and carried out 24/7 by computers which never sleep. In this sense, Masdar City itself is a laboratory which never sleeps.

Masdar City is also a showroom which never closes. While indoor laboratories are seldom fully and always open, this is not the case for cities. The emirate's new urban settlement does not have gates or office hours. People can enter Masdar City and walk around at any time. Given that, as part of the Masdarian deal, the companies in partnership with the Masdar Initiative install their products in the new city, most of their creations are constantly visible. Therefore, potential buyers, developers and investors can observe the portfolio of the Masdar Initiative's partners by simply walking around Masdar City. New technologies are everywhere for everybody to be seen. They permeate the built environment, at times becoming almost indistinguishable from it (see, for instance, Figure 4.1). 'We want to bring as much portfolio as possible to showcase,' explained a marketing manager from Siemens, in relation to the choice of the German company to build its Middle-East headquarters in Masdar City.

After being developed, tested and showcased in Masdar City, Masdarian technologies are commercialized and sold worldwide. 'We test all our products in Masdar City. Then, if they are successful, we roll them out globally' a Siemens spokesman stated, specifying that 'the objective is mass production.' The circle is closed by splitting the revenues that the sale of each product generates. When companies like Siemens and Schneider sell a product which was developed and tested in Masdar City, they have to share part of the profit with the Masdar Initiative. The percentage varies from product to product, depending on the shares that the Masdar Initiative owns. As a member of the Masdar Initiative specified in an interview,

> if it is a local partner from the GCC Area (Gulf Cooperation Council), Masdar will be the majority shareholder, meaning 60 per cent to 40 per cent ownership and profit. If the development of a new technology is conducted with an international partner, the ownership will vary from 30 per cent to 40 per cent.

FIGURE 4.1 Roof-mounted solar panels in the city centre of Masdar City.
Source: Antonio Mannu

The business package that the Masdar Initiative offers does not end here. The promotion and commercialization of the products made in Masdar City are supported
by a web portal called The Future Build (TFB). Launched in March 2010 by the
Masdar Initiative, TFB (2020) is an international platform functioning as an online
store for cleantech products. The design mirrors e-commerce sites *à la* Amazon. Here
the offer of a product comes with related information and testimonials from multiple
third parties (mostly clients from all over the world) and, above all, by Masdar City
itself. In TFB, Masdar City is portrayed as an entity whose presumed authority on
urban sustainability can certify the quality of the technologies that bear its label.
According to the narrative of TFB, engineers, developers, contractors and architects
can find everything that is needed to 'build sustainability,' by following in the
footsteps of 'one of the most sustainable, visually stunning and liveable developments
on earth' (TFB, 2020: no page). In practical terms, the partners of the Masdar Initiative
can use TFB to display their portfolio, showing that what they offer has been tested in
Masdar City. 'I want to use the same aluminium they are using in Masdar City,' said a
representative from the Masdar Initiative in an interview, to exemplify the attitude of
potential clients shopping online in the Masdarian digital marketplace.

In addition, the Masdarian business is supported by international events such as the
World Future Energy Summit and the European Future Energy Forum, organized by
the Masdar Initiative every year. While the geographical location varies, the formula
does not. The World Future Energy Summit (2020), for instance, is publicized as
'a global industry platform connecting business and innovation in energy, clean

technology and efficiency for a sustainable future.' In practise, it is a bombastic exhibition where clean technologies are showcased. During the event, renowned personalities invited by the Abu Dhabi Government, like Ban Ki-moon and Hilary Clinton, give talks about global environmental problems. Conveniently, visitors and participants are surrounded, much like at a trade fair, by stands where clean-tech companies claim to have the solution to those very problems. The solution bears the label of Masdar City, which is promoted as 'a model for how we'll live tomorrow' (see Figure 4.2).

The second mechanism through which the Masdar City project feeds into Vision 2030 and the development of a post-oil economy, pivots around intellectual property gains. The products co-developed by the Masdar Initiative and its partners come with intellectual property rights such as patents and industrial design rights. These types of rights are investable assets which investors and brokerage firms can buy, betting on the success of the Masdarian products. In addition, other companies can purchase them to produce and commercialize a Masdarian product themselves, or to extract royalties from rival companies which infringe a patent. To put the potential of this business mechanism into perspective, it is important to remember that a large share of the portfolio of the technologies made in Masdar City is related to photovoltaics and solar power. Global investment in renewable energy is

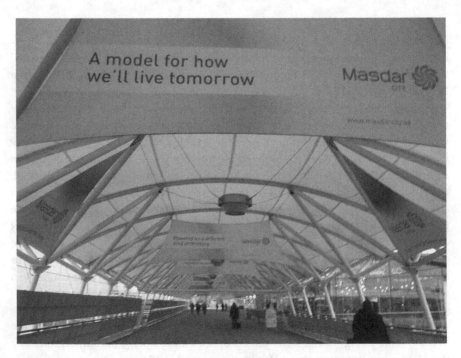

FIGURE 4.2 Promotion of Masdar City at the 2010 European Future Energy Forum in London.

Source: Author's original

growing and solar accounts for the majority of it (Frankfurt School-UNEP Centre/ BNEF, 2018). In 2017, investments in solar energy technologies overshadowed all the investments in both renewable and non-renewable energy, with China alone reaching the record figure of US$ 86.5 billion (Frankfurt School-UNEP Centre/ BNEF, ibid.). This is the market to which a substantial part of Masdar City's business is oriented.

Last but not least, there is the business of real estate. The partnerships between the Masdar Initiative and international cleantech companies do not lead only to commercial products. Partnerships also lead to rents. The partners of the Masdar Initiative move to Abu Dhabi's new city, establishing new offices, laboratories and showrooms. In so doing, they claim a portion of Masdar City which is not a free space. The government of Abu Dhabi owns Masdar City and the land upon which it sits. Therefore, if a company wants to be based in Masdar City, it has to pay. In practise, the Abu Dhabi Government acts as a landlord, and the Masdar Initiative as the real-estate agent that sets and collects the rent. Renting costs depend largely on how much space a company wants to occupy and where that space is located, with charges rising near the city centre. As a manager from Schneider declared in an interview, 'here we are simply tenants.' To date, the Masdar City project is the largest and most expensive urban project developed in Abu Dhabi. According to a representative from the Masdar Initiative, once the construction of the new city is completed, the profit generated by its lease is expected to overcome that of any other property in the emirate.

The planning of Masdar City

The business underpinning Masdar City has profound implications for the way the new city is planned and designed. These are tasks for which F+P, with its team of urban planners and urban designers temporarily based in Abu Dhabi, is officially responsible. However, as this section illustrates, the Masdarian business considerably complicates the planning process as well as the nature of roles and responsibilities. Abu Dhabi's eco-city project exists to support the implementation of Vision 2030 by means of strategic partnerships with cleantech companies. Without these partnerships there is no reason for Masdar City to exist, and therefore there is no reason to plan its construction. As a leading planner from F+P explained in an interview, 'Masdar City is a demand-responsive development' which does not go ahead until there are companies ready to move in. Since the Emirati city is being built from scratch and its final form was visualized through a master plan, a common assumption is that the developers are following a precise programme in a systematic manner. In reality, this is far from the case. The initiative is not in the hands of F+P's planners: it lies in the agenda of the managers that control the Masdarian business.

The private companies in partnership with the Masdar Initiative have the power to steer the planning process. They can choose what portion of the new city will be theirs, what will be built in their part of Masdar City and also what their space will look like. Companies such as Siemens, Schneider and Mitsubishi are given the opportunity to establish their headquarters or a branch in the Emirati settlement, and to impose their imagination over the master plan of F+P. A spokesman of the Masdar Initiative explained in an interview that many multinationals have their own ideas of space, as well as an aesthetic which they want to see reflected in all their buildings, regardless of the geographical location. The Masdar City project accommodates this trend. The business partners of the Masdar Initiative keep their philosophy of design when they move to Masdar City by bringing their own architects. As a manager from Siemens stated, 'we have our own architects.' He clarified that F+P is not in charge of the design of Siemens's spaces. Siemens is. In practise, the architects and designers paid by Siemens conceptualize and draw the building of the German multinational in an independent manner. When Siemens is satisfied with their work, it sends a construction plan to the Masdar Initiative which, in turn, sends it to F+P whose job is to modify the master plan according to Siemens's plan. Ultimately, the overall impact on Masdar City can be substantial. In 2014, for example, Siemens completed the construction of its Middle-East headquarters: a large office building which occupies 20,000 m^2, forming a block of the new city.

As a planner from F+P confessed in an interview, 'the process is frustrating.' The business partners of the Masdar Initiative have the last word over their buildings and, from a planning point of view, the construction plan of a single company carries more weight than the whole master plan. The latter adapts to the former, accommodating the will of companies like Siemens and Mitsubishi which do not have the slightest obligation towards the directions set by Foster's planners. F+P has to keep the master plan flexible, aware that every time the Masdar Initiative signs a partnership, the Masdar City project will change. On these terms, F+P's master plan serves the purpose of tracking the constant transformations of Masdar City, instead of disciplining its physical development in a coherent way. 'In the beginning of the project, we tried to establish a set of key principles' recalled a member of Foster's studio reflecting on the evolution of the master plan of Masdar City. 'For example, we wanted and tried to get rid of cars. But now, as you will see, these principles are being ignored' he concluded, blaming the Masdar Initiative for prioritizing the needs of its partners over F+P's ideas.

F+P also has very little control over the temporality of the project. If there is no new partnership coming up, meaning that there is no new company moving in, the planning process stops. Construction work stops too. The parts of Masdar City already built and functional continue their activities normally, while the rest of the city is forced to enter an artificial coma (Cugurullo, 2016b). Only when a new partnership is signed and fresh capital is pumped into the project, the planning machine is set in motion again. What happens next is unknown to F+P's planners. It will be the job of the architects, brought in by the latest partner of the Masdar Initiative, to design the latest building in Masdar City. F+P will metabolize the new external inputs, no matter how different they are from their original vision. In the words of a leading planner from F+P, the main task of his company is not to plan the development of new buildings and infrastructures, but rather to find a way to 'plug' the spaces envisioned by the Masdar Initiative's partners into Masdar City (Cugurullo, 2016c).

The Masdarian economics have a deep influence not only on the planning of the new city, but also on its physical shape and structure. A key business mechanism of the Masdar City project allows private companies to install their products in the built environment. In so doing, the partners of the Masdar Initiative change the materiality of Masdar City, which evolves to accommodate their portfolios. If a company like Siemens, for instance, decides to invest in roof-mounted solar panels, this means that some of the roofs of the buildings in Masdar City will be designed and then constructed to feature solar panels. If the same company later decides to invest in the development of a smart grid, the very ground of Masdar City will be reshaped to allow the passage of cables and fibre channels. If the latest Masdarian technology consists of a concentrated solar power station, the material consequence of this business strategy will be the development of a whole new urban area.

It is important to remember that the Masdar City project includes several different companies which invest in diverse clean technologies. Moreover, even the same type of clean technology, such as photovoltaics, for example, can come in many different shapes and sizes. Consequently, Masdar City accommodates a variety of devices whose heterogeneous materiality leads to the production of a variety of spaces. The spectrum of the Masdarian technology ranges from micro smart sensors to 20 meter-high solar power towers (see Figure 4.3). There is no physical or conceptual limit to this spectrum. The sky is not the limit. The business is the limit. As long as the Masdar Initiative sees profit in an experimental technology, its size and shape do not matter. What is deemed profitable gets built. Market analyses and business strategies lead, and urban planning and design follow.

What is crucial to note is that, even if F+P keeps the master plan flexible to make every new Masdarian technology compatible with Masdar City, the Masdarian technologies are not necessarily compatible with each other. To understand this problem, the first step is to see Masdar City not as a single and homogeneous project, but rather as the sum of different sub-projects. As discussed in the previous section, every time the Masdar Initiative forges a partnership with a cleantech company, the two companies work together on the development of a new technology which will be eventually realized and integrated into Masdar City, to be tested and showcased. The development of a new Masdarian product is therefore a project per se, which starts in the office where the partnership is signed, and ends in the part of the new city where the technology is installed. Every technological project is also an urban project, inasmuch as it shapes the built environment. Seen from this perspective, Masdar City is a space where different independent projects are constantly being developed and implemented, without a rigid master plan operating as a filter. 'We do not have a framework to follow' made clear in an interview a representative from Siemens, with regard to the development of the buildings of the German company. As a result, by planning the construction of diverse technologies, the many companies in partnership with the Masdar Initiative are planning the construction of Masdar City. F+P then is not *de facto* in charge of the planning process.

FIGURE 4.3 A beam-down solar power tower in Masdar City.
Source: Gianfranco Serra Photography

Under these conditions, clashes and tensions among individual cleantech projects are inevitable. An example is the development of Personal Rapid Transit (PRT) and the development of electric cars: both occurring in Masdar City. The PRT project was ideated by F+P and is part of the original master plan of Abu Dhabi's alleged eco-city. It is a system of driverless automated vehicles for up to four passengers meant to eliminate automobile dependency and decrease energy consumption (see Figure 4.4). A key conceptual underpinning of the PRT project is the philosophy of sharing. The inhabitants of Masdar City do not own a PRT vehicle. They share one whenever they want, for free, as any person visiting Masdar City would. In this sense, the PRT project was designed as a large-scale, free

FIGURE 4.4 A PRT driverless car in Masdar City's undercroft. The automated vehicle waits for its next passenger in an underground station.
Source: Antonio Mannu

public transport service. Running in parallel, however, is another urban transport project carried out by Mitsubishi, a key partner of the Masdar Initiative, which is based on an oppositional philosophy and target. Mitsubishi's goal is to develop and commercialize electric cars. The technology in question is a private means of transport. It cannot be used without first being purchased. This project targets the individual rather than the community and does not contemplate how and the extent to which the driver will use the vehicle. On the one hand, there are PRT vehicles which are automated and follow sensors integrated into the built environment, to connect every area of the city via the shortest route to save energy (Cugurullo and Acheampong, 2020). On the other hand, there are the vehicles offered by Mitsubishi which are not geographically restrained, and place the responsibility for the route and its energy cost on the driver. Moreover, the PRT project, with its focus on sharing, was intended to create a mobile public space for social encounters. Instead, the Mitsubishi project focuses on private property, private time and, ultimately, private space.

In addition to not being in harmony for a conceptual point of view, the two projects are incompatible in urban design and infrastructure. The PRT system is engineered to function mainly underground. F+P's initial master plan pictured Masdar City divided into two levels. Above the ground, pedestrian spaces consisting mostly of narrow alleys, and not a single car. Below, an undercroft for the circulation of PRT vehicles. This urban design was also meant to maximize the amount of green space on the surface. Originally, the new Emirati city was supposed to be crossed transversally and horizontally by three narrow parks whose purpose was to channel the winds of the desert and decrease the perceived temperature in the city (for the original rendering, see Foster + Partners, 2007). Mitsubishi's cars cannot operate in this type of built environment. An average car needs a street which is broader than those used by pedestrians, and its mobility would be severely undermined in a narrow alley full of people. Furthermore, a functional system of urban transport based on cars requires an extensive network of roads crisscrossing the entire city, and cannot coexist with the continuous lines of green spaces designed by F+P. Such inner tensions make the Masdar City project intrinsically dysfunctional and prone to fragmentation. Although from a business perspective, the Masdar Initiative wants and, to some extent, can keep the Masdar City project flexible and open to new sub-projects, Masdar City as a physical environment is ultimately a finite space. Because of these unavoidable physical limitations, choices about what should be built and kept have to be eventually made: choices that have repercussions on the sustainability performance of the Emirati eco-city project.

The sustainability of Masdar City

The economic focus of the Masdar City project has a profound influence on the Masdarian understanding and practise of sustainability. When asked about the interpretation of the idea of sustainability in the Emirati eco-city project, a manager from the Masdar Initiative asserted: 'Here we are focusing on making sustainability commercial.' One major implication of this business-oriented notion of sustainability is the fact that the developers evaluate the success of the project and the performance of the city largely from an economic point of view. In this regard, the case of the PRT is emblematic. Although successful from an environmental and social perspective, the PRT project was officially abandoned in October 2010. Environmentally, Masdar City's system of automated vehicles was saving energy and decreasing the carbon emissions produced by the settlement. Socially, it was reducing automobile dependency and encouraging people to walk, thus cultivating social encounters in the city (as well as inside the PRT vehicles). Economically, however, the Masdar Initiative was not generating profit out of the PRT. This single economic negligence was enough for the developers to decree the end of the PRT project. As a manager from the Masdar Initiative put it in an interview, 'the PRT costs a fortune and Masdar City is not an environmental crusade.' The Emirati public company did not see in the PRT project a strong return on investment, and decided to stop its implementation and sign a partnership with Mitsubishi to co-develop electric cars.

The moment the development of the PRT project stopped, the development of Masdar City as a car-free city stopped too. When this happened, approximately 10 per cent of the original master plan signed by Foster had already been implemented. Today, this percentage consists of the city centre of the new Emirati city: the only area where the PRT is functional and streets are fully pedestrian. This is also the only part of Masdar City where the initial vision of ecological urbanism cultivated by F+P can be observed. In order to achieve sustainability, F+P's planners and architects had taken inspiration from traditional Islamic architecture, 'because if you want to build the city of the future, it would be arrogant or ignorant to design everything new. You need to look into what has been achieved in the past and what can still be applied' stated a member of F+P in an interview. More specifically, Aleppo (Syria), Fez (Morocco) and Shibam (Yemen), due to their compact urban design, had been used by Foster's studio as models, with the aim of maximizing passive shading and air circulation. This idea was successfully realized. The city centre of Masdar City is characterized by narrow streets where the proximity of the buildings leads to natural shading and channels the winds of the desert. As a result, in this built environment, the perceived temperature can be up to 15 degrees lower than in an average street of Abu Dhabi. In sustainability terms, the difference is significant. In summer, for example, this is the difference between 35°C and 50°C which, in turn, can be the difference between opting for walking instead of driving. Indoors, this 15-degree difference may mean that air-conditioning is only turned on occasionally, rather than being permanently operational. However, although environmentally and socially sustainable, in the eyes of the managers of the Masdar City project, this type of urban design was not economically profitable. A pedestrian street is not a commodity and, as such, it cannot be sold. A car, on the other hand, is a commodity and it can be sold worldwide. This is the reason why, since 2009, Masdar City stopped being a laboratory for alternative urban mobilities, and embraced a traditional car-friendly layout to support the Masdar Initiative's partnerships with car manufactures. The undercroft, initially developed to separate people from means of transports, was never completed, and it now resembles a crypt where the few remaining PRT vehicles still function.

The economics-driven understanding and practise of sustainability that permeates Masdar City is a reflection of Vision 2030 and its *sustainability is profitability* mantra. One Masdar Initiative manager repeated several times in an interview that the main purpose of Masdar City is to 'make sustainability commercial,' specifying that profitability is a *conditio sine qua non*. He argued that 'Masdar City should not be treated as a charity' because 'it has money given to it by the government, and the government expects to see a return on investment.' Emblematically, during the whole interview, the manager kept referring to Masdar City not as a city, but as a product. 'If you want to make a product,' he stated, 'the first question you want to ask is: is there a demand for it?' For him, Masdar City is a successful story, because the technologies and services that the project offers are in demand, and the Masdar Initiative and its partners have clients from all over the world. When asked about the extent to which Masdar City is sustainable, the manager replied by saying that the new city is sustainable because it is 'economically viable' and 'cost-effective.' 'I am paying my cost and I am making money' he concluded, summarizing the essence of the Masdarian philosophy of sustainability.

However, the fact that the Masdar City project has shown to have the potential of being economically profitable over a long period of time, does not mean that the new Emirati city is sustainable. There are other dimensions to take into account. The environmental dimension of Abu Dhabi's eco-city project, for instance, presents three severe lacunae. First, while the focus of the developers is to quickly build an 'unmatched platform for the commercial-scale demonstration of sustainable technology' and sell Masdarian products globally, little consideration is given to the means that serve the end (Masdar Initiative, 2011: no page). The construction of such a platform implies the construction of a brand-new urban settlement in the desert: an energy-intensive process which requires the consumption of a large amount of resources. From an energy point of view, Masdar City is not built on sand. It is built on oil, and black is the colour that sustains this alleged green city. As one F+P planner admitted in an interview, while Masdar City boasts a plethora of generators of renewable energy, it will be able to generate enough energy to sustain itself only after its completion. In essence, this means over two decades of fossil-fuel dependency. In addition, since Abu Dhabi has scarce fresh water resources, Masdar City relies and will keep relying on desalination: a need which requires oil-powered facilities and that will grow together with the city. This problem is even more severe because, as an engineer from the Masdar Initiative revealed, 'Masdar City cannot even have its own desalination plant. The water nearby is too salty.' As a result, the life of the city depends on a complex and extended energy-intensive infrastructure which mobilizes vast off-site resources, channelling them into Masdar City.

Second, the Masdar City project has a narrow interpretation of environmentalism. The environmental attention of the developers is almost exclusively on those environmental problems where the solution can be capitalized on. In terms of environmental sustainability, the technologies that are developed and installed in Masdar City tackle the reduction of carbon emissions. Roof-mounted solar panels, smart grids, electric cars and concentrated solar power stations, despite the sheer diversity in terms of design and engineering, have a common denominator: CO_2. The Masdarian technology focuses on curbing CO_2 emissions, because this type of technology is in demand. Behind the development of the environmental dimension of Masdar City are not environmental analyses showing what environmental problems urbanization is responsible for, but rather market analyses showing what environmental problems can be easily monetized. It is scientifically established that urban carbon emissions contribute to global environmental problems such as climate change (Bai et al., 2018). Therefore, the technological arsenal of Masdar City can be, to some extent, environmentally beneficial. However, as discussed in Chapter 1, it is equally established that urbanization is causing a number of other urgent environmental issues which include water scarcity, loss of natural habitat and the destruction of ecosystems and ecosystem services. This spectrum of ecological problems is not taken into account by the lucrative narrative of Abu Dhabi's eco-city project, inasmuch as it is not easy to turn into profit.

Third, in Masdar City, *environmentalism* is understood as *consumerism*. While it professes the decarbonization of cities, the Masdarian narrative of sustainability does not question fundamental issues of consumption. On the contrary, it encourages consumption provided that what is being consumed are the products tested and developed in Masdar City. The story told by the Masdar Initiative does not mention that its clean technologies are ultimately objects whose creation requires intense processes of extraction and production. Electric cars, smart grids and photovoltaics necessitate materials such as metals, minerals and plastics, which are extracted from the ground or derived from petroleum. In addition to exploiting the stock of resources of the planet, the extraction process is carbon intensive and has an immediate negative impact on local ecosystems. Furthermore, it adds to the carbon emissions produced when materials are assembled into products, and when products, in the shape of commodities, are distributed across the world. Essentially, the business at the core of Abu Dhabi's eco-city project, crafts a narrative of environmentalism which hides the negative environmental side of what is being promoted as *eco-friendly*. The narrative is also carefully crafted to avoid any end, inasmuch as this would imply the end of the Masdarian business. The story of consumption narrated by the Masdar Initiative has neither a temporal limit, nor a geographical boundary. According to it, people can continue to consume and cities can keep growing *ad libitum*.

Similarly, the social dimension of Masdar City presents major lacunae caused by the pervasive influence of the Masdarian business. As a manager from the Masdar Initiative argued in an interview, 'Masdar City must be attractive.' The project has to attract private companies to Abu Dhabi, and convince them not simply to co-develop new cleantech products with the Masdar Initiative, but also to move their offices to the emirate. For this reason, when questioned about how social sustainability is understood and practised in Masdar City, the manager replied: 'customer satisfaction.' On these terms, the Masdarian clientele is twofold. First, the partners of the Masdar Initiative are the customers. Before signing a partnership, companies such as Siemens and Mitsubishi can visit Masdar City and see with their own eyes how the project functions. They can then select the portion of the settlement that they want to occupy and even bring their own architects to design it. Nothing is hidden. Prior to relocating, potential partners spend months in Abu Dhabi, visiting the construction site and learning about what facilities and services they would have available by signing the deal offered by the Masdar Initiative. When the conditions are fully clear, partnerships are signed. Second, the clients of the partners of the Masdar Initiative are the customers. The buyers of the clean technologies tested and developed in Masdar City can see how these products operate in a real-life environment before finalizing their purchase. Through The Future Built, they can even go online and check the ratings and the reviews of the Masdarian portfolio. This is what the manager from the Masdar Initiative described as 'fairness': the advantage of knowing what you are buying. However, the Masdarian fairness is only economic in nature and, with its narrow understanding of sustainability, it excludes pressing issues of social and environmental justice.

The exception that proves the rule is what the Masdar Initiative calls the *20 per cent policy*. Echoing ideals of social justice and equity common in sustainability discourses, this policy reserves 20 per cent of the housing space of the new city for low-income workers. However, in addition to the fact that it comprises only a small percentage of Masdar City, the '20 per cent policy' is severely under-developed. As a planner from F+P remarked in an interview, there is no clear definition of 'low-income worker' and, above all, priority has been given to the construction of high-standard accommodations. The '20 per cent policy' might be implemented at a later stage which was unknown to all the members of F+P that were interviewed, including the studio's chief planner. As of this writing, renting a standard one-bedroom apartment (65 m^2) in Masdar City costs at least €1250 monthly, while the majority of the population of Abu Dhabi (consisting of low-income foreign workers) has an average salary of €220 per month, despite working 12 hours a day, seven days a week. Masdar City is not for everybody. Nor is the profit that its business generates. The capital produced by the new city immediately leaves the settlement to feed directly into the wealth of the Emirati royal family, and indirectly into the wealth of the Emirati citizens through the national welfare system. The rest of the population is left, physically and eco-nomically, out of the city.

Conclusions

Although officially pictured as a standalone city in the desert, Masdar City is intrinsically part of Abu Dhabi, and its genesis cannot be understood without understanding its geographical context. As this chapter has shown, the story of the development of Masdar City is part of a broader story of regional development. The Emirati project for a new eco-city has been launched in line with Abu Dhabi's new development agenda, Vision 2030, whose purpose is to diversify the local economy, by establishing a post-petroleum economic sector. There is a symbiotic relationship between Vision 2030 and Masdar City, for the former feeds off the latter. Vision 2030 establishes an ideal political economy, and the Masdar City project is one of the main tools used by the Abu Dhabi Government to realize it. Moreover, conceptually, the Masdarian philosophy of sustainability derives from the politico-economic targets and priorities set by the Emirati policy-makers in Vision 2030. In this sense, Masdar City approaches sustainability mostly in eco-nomic terms, because the Masdar City project itself is the outcome of a policy agenda which approaches development primarily from an economic perspective.

At the core of the Masdar City project is a business crafted to achieve the post-petroleum political economy envisioned by the sheikh. Its mechanics are care-fully designed and implemented to generate capital, without relying directly upon oil exports. The government of Abu Dhabi is aware that the emirate is running out of oil and that alternative sources of capital are necessary in order to preserve the status quo. In this context, the contribution of the Emirati eco-city project is threefold. First, the new city is developed as a living laboratory where cleantech multinationals co-develop with the Masdar Initiative experimental technologies which are installed and showcased in Masdar City, to be eventually sold worldwide. On these terms, Abu Dhabi's eco-city project is simultaneously an office, a laboratory, a factory, a showroom and a shop. Second, all the Masdarian products are associated with investable assets which are part of one of the fastest-growing markets in the world (solar energy). In this regard, Masdar City is a magnet for global investments. Third, the new settlement is a large space which the Abu Dhabi Government rents out, attracting companies such as Siemens and Mitsubishi as tenants. Seen from this perspective, the Emirati project for an eco-city is, in practise, a huge rental property. Through these three business mechanisms, the Masdar City project also triggers the installation of generators of renewable energy, which tackle important regional issues of energy security. Roof-mounted solar panels, solar farms and futuristic solar towers are a common sight in Masdar City. However, what this clean technology produces the most is not non-oil energy. It is non-oil capital. As this chapter has highlighted, the new city is far from producing enough energy to sustain even itself, and it must rely on fossil fuels on a daily basis. The business, on the other hand, is bearing fruit. Multinationals are currently renting out large segments of Masdar City, using their space to create and commercialize products which are sold all around the world.

The empirical exploration of Masdar City has revealed that the Emirati eco-city project, due to its economic imperatives, is in sum an international shop, a laboratory, an office, a factory, a showroom, a magnet for global investments and a rental property. This picture also tells what Masdar City is not. Put simply, the new settlement lacks a social dimension. The project does not include a single initiative meant to cultivate a society in Masdar City. Moreover, the developers, in the pursuit of profit, are willing to sacrifice elements of the project, such as pedestrian public spaces and shared transport systems, which could help the new city grow a society. Consequently, Masdar City does not have a society and, without this essential social component, it cannot be called a city. The very individuals who operate the Masdarian business do not see Masdar City as a place to live, or just to occasionally experience for reasons that are not connected to economic activities. In an interview, a manager from Schneider was asked why his company was based in Masdar City. His reply was: 'We are here to live what we preach.' This answer is fundamentally flawed. Not a single person working for Schneider is living in Masdar City. Likewise, no one from Siemens, Mitsubishi or any other company in partnership with the Masdar Initiative is living in the new city. The members of the Masdar Initiative themselves do not live in Masdar City. They all commute. They live in other parts of Abu Dhabi or in Dubai. Some of them even live outside the UAE and enjoy Masdar City's strategic location next to Abu Dhabi International Airport. For them, Masdar City is a space to work, not a place to live.

Today, after over a decade since the beginning of the project, less than 1000 people live in Masdar City. Its population consists largely of masters students and PhD researchers (funded by the Masdar Initiative and its business partners) who work full-time on the development of new Masdarian technologies. Because of the extremely low number of actual residents and the fact that the original goal of the developers was to attract 50,000 people, Masdar City can be seen as a ghost town and, indeed, this is what it has been often called in the media (see, for instance, South China Morning Post, 2018; The Guardian, 2016). However, this terminology does not fully expose the core problems of Masdar City and the dangers that the Masdarian model of city-making can produce. A ghost town implies the presence of the spectre of a community that once was. This is not the case of Masdar City. From a social point of view, the new Emirati city is not undead. It is unborn. As a representative from the Masdar Initiative put it in an interview, 'Masdar City at the end of the day is a business.' Such existence funded upon a project of capital accumulation is precarious. The very foundations of Masdar City are unstable. As repeatedly pointed out by Harvey (2010), capital is unstable, volatile and its mutations and movements are often unpredictable. Above all, capital does not bother to fix crises. It moves past them, following new geographical trajectories, in the perpetual search of more profitable venues. If the Masdarian business declined, Masdar City would collapse. A city without social foundations is an unsustainable non-place (Augé, 1995). Systemic crises are part of capitalism and only strong and cohesive societies can endure them.

The contemporary health of the market where the Masdarian business is situated is misleading, and it can obfuscate the actual capacity of the Masdar City project to sustain itself. In addition to the fact that there is no guarantee that global investments in the type of technology co-developed by the Masdar Initiative and its partners will keep rising, a growing market means a growing competition. In this regard, Abu Dhabi competes against other powerful countries which are investing in the production of the same clean technologies, by means of national companies and international partnerships akin to the Masdarian model. This is particularly the case of China which in 2017 invested US$ 126.6 billion in renewable energy (mostly solar power), while the United Arab Emirates' total investment was much less: US$ 2.2 billion (see Frankfurt School-UNEP Centre/BNEF, 2018). What might also happen is that, in the near future, some of the business mechanisms examined in this chapter will be outdated or ineffective. Partnerships might break, e-commerce sites might close, and trade fairs might disappear. Should this happen, Masdar City would lose its foundations and break down. No society would prevent or lament the fall. There are people in Masdar City and the total population is likely to grow, but the mere concentration of human beings in a given geographical space does not make a society, just like the sole concentration of capital is not enough to forge a city. Rather than an empty ghost town, Masdar City is a sandcastle. It has a magnificent appearance, but the structure is fragile. It is filled and full, but its substance and mass are like sand. Masdar City would crumble at the first wave of economic crisis.

There is a stark difference between the theories of ecological urbanism, examined in Chapter 2, and the Masdarian ideas and practises of sustainability. In ecological urbanism, *eco* stands for *ecological*. In Masdar City, *eco* stands for *economic*. Business and profitability are the core elements of the Masdar Initiative's formula for sustainable urban development. What is not profitable is left out of the equation: a formulation which severely undermines the sustainability of Abu Dhabi's new city. Crucial environmental themes, such as water scarcity and ecosystem services are ignored. Pressing social questions of justice and equity are unanswered. Moreover, while the Emirati alleged eco-city, with its arsenal of clean technologies can reduce carbon emissions, the production and consumption of the Masdarian technology generate carbon emissions in the first place. In this sense, the formula developed by the Masdar Initiative to create sustainability, is the same formula that produces unsustainability. At the end of the story of the Masdar City project, the hero supposed to bring sustainability to all the cities in the world, turns out to be the villain.

References

Abu Dhabi Government (2008). Economic Vision 2030. [Online] Available: http://gsec. abudhabi.ae/Sites/GSEC/Content/EN/PDF/Publications/economic-vision-2030-full-version,property=pdf.pdf [Accessed 10 November 2020].

Abu Dhabi Urban Planning Council (2008). *Urban Planning Vision*. UPC Press, Abu Dhabi.

Abu Dhabi Urban Planning Council (2013). *Urban Structure Framework Plan*. [Online] Available: http://www.upc.gov.ae/abu-dhabi-2030/capital-2030.aspx?lang=en-US [Accessed 10 November 2020].

Ali, S. (2010). *Dubai: Gilded cage*. Yale University Press, New Haven.

Augé, M. (1995). *Non-places: Introduction to an Anthropology of Supermodernity*. Verso, London.

Bai, X., Dawson, R. J., Ürge-Vorsatz, D., Delgado, G. C., Barau, A. S., Dhakal, S., Dodman, D., Leonardsen, L., Masson-Delmotte, V., Roberts, D. and Schultz, S. (2018). Six research priorities for cities and climate change. *Nature*, 555 (7694), 23–25.

Bibri, S. E. (2020). The eco-city and its core environmental dimension of sustainability: green energy technologies and their integration with data-driven smart solutions. *Energy Informatics*, 3 (1), 1–26.

Caprotti, F. (2014a). *Eco-cities and the transition to low carbon economies*. Springer, Berlin.

Caprotti, F. (2014b). Critical research on eco-cities? A walk through the Sino-Singapore Tianjin Eco-City, China. *Cities*, 36, 10–17.

Chang, I. C. C. and Sheppard, E. (2013). China's eco-cities as variegated1 urban sustainability: Dongtan eco-city and Chongming eco-island. *Journal of Urban Technology*, 20 (1), 57–75.

Chang, I. C. C., Leitner, H. and Sheppard, E. (2016). A green leap forward? Eco-state restructuring and the Tianjin–Binhai eco-city model. *Regional Studies*, 50 (6), 929–943.

Chapman, I. (2014) The end of Peak Oil? Why this topic is still relevant despite recent denials. *Energy Policy*, 64, 93–101.

Cugurullo, F. (2013a). How to build a sandcastle: An analysis of the genesis and development of Masdar City. *Journal of Urban Technology*, 20 (1), 23–37.

Cugurullo, F. (2013b). The Business of Utopia: Estidama and the Road to the Sustainable City. *Utopian Studies*, 24 (1), 66–88.

Cugurullo, F. (2016a). Urban eco-modernisation and the policy context of new eco-city projects: Where Masdar City fails and why. *Urban Studies*, 53 (11), 2417–2433.

Cugurullo, F. (2016b). Speed kills: fast urbanism and endangered sustainability in the Masdar City project. In Datta, A. and Shaban, A. (Eds.) *Mega-urbanization in the Global South: Fast cities and new urban utopias of the postcolonial state*. London: Routledge, pp. 78–92.

Cugurullo, F. (2016c). Frankenstein cities. In Evans, J., Karvonen, A. and Raven, R. (Eds.) *The experimental city*. Routledge, London.

Cugurullo, F. (2018). Exposing smart cities and eco-cities: Frankenstein urbanism and the sustainability challenges of the experimental city. *Environment and Planning A: Economy and Space*, 50 (1), 73–92.

Cugurullo, F. (2020). Urban Artificial Intelligence: From Automation to Autonomy in the Smart City. *Frontiers in Sustainable Cities*, 2, 38. doi:10.3389/frsc.2020.00038.

Cugurullo, F. and Acheampong, R. (2020). Smart cities. In Jensen, O., Lassen, C., Kausfmann, V., Freudendal-Pedersen, M. and Lange, I.S.G. (Eds.) *Handbook of Urban Mobilities*. London: Routledge, pp. 389–297.

Cugurullo, F. and Ponzini, D. (2018). The transnational smart city as urban eco-modernisation. In Karvonen, A., Cugurullo, F. and Caprotti, F. (Eds.) *Inside Smart Cities: Place, Politics and Urban Innovation*. Routledge, London.

Davidson, C.M. (2009). *Abu Dhabi: Oil and beyond*. C. Hurst & Co., London.

De Jong, M., Joss, S., Schraven, D., Zhan, C. and Weijnen, M. (2015). Sustainable–smart–resilient–low carbon–eco–knowledge cities; making sense of a multitude of concepts promoting sustainable urbanization. *Journal of Cleaner Production*, 109, 25–38.

De Jong, M., Yu, C., Joss, S., Wennersten, R., Yu, L., Zhang, X. and Ma, X. (2016). Eco city development in China: addressing the policy implementation challenge. *Journal of Cleaner Production*, 134, 31–41.

De Jong, M., Hoppe, T. and Noori, N. (2019). City Branding, Sustainable Urban Development and the Rentier State. How do Qatar, Abu Dhabi and Dubai present Themselves in the Age of Post Oil and Global Warming?. *Energies*, 12 (9), 1657.

Foster + Partners (2007). Masdar development. [Online] Available: http://www.foster andpartners.com/projects/masdar-development/ [Accessed 10 November 2020].

Frankfurt School-UNEP Centre/BNEF (2018). Global trends in renewable energy investment 2018. [Online] Available: http://www.iberglobal.com/files/2018/renewable_trends.pdf [Accessed 10 November 2020].

Griffiths, S. and Sovacool, B. K. (2020). Rethinking the future low-carbon city: Carbon neutrality, green design, and sustainability tensions in the making of Masdar City. *Energy Research & Social Science*, 62, 101368.

Harvey, D. (2010). *The enigma of capital*. Profile Books, London.

International Monetary Fund (2013). Report for selected countries and subjects. [Online] Available: http://www.imf.org/external/data.htm [Accessed 10 November 2020].

Joss, S., Cowley, R. and Tomozeiu, D. (2013). Towards the 'ubiquitous eco-city': an analysis of the internationalisation of eco-city policy and practice. *Urban Research & Practice*, 6 (1), 54–74.

Lin, G. C. and Kao, S. Y. (2020). Contesting Eco-Urbanism from Below: The Construction of 'Zero-Waste Neighborhoods' in Chinese Cities. *International Journal of Urban and Regional Research*, 44 (1), 72–89.

Linz, J. J. and Stepan, A. C. (1996). *Problems of democratic transition and consolidation: southern Europe, South America, and post-communist Europe*. Johns Hopkins University Press, Baltimore.

Luomi, M. (2009). Abu Dhabi's alternative-energy initiatives: Seizing climate-change opportunities. *Middle East Policy*, 16 (4), 102–117.

Masdar Initiative (2011). Masdar City. [Online] Available: http://www.masdarcity.ae/en/ 27/what-is-masdar-city-/ [Accessed 10 November 2020].

Masdar Initiative (2012). *The reality of future energy*. Masdar Initiative, Abu Dhabi.

Masdar Initiative (2018). About Masdar City. [Online] Available: http://www.masdar.ae/ assets/downloads/content/270/masdar_city_brochure.pdf [Accessed 10 November 2020].

McCann, E. (2011). Urban policy mobilities and global circuits of knowledge: Toward a research agenda. *Annals of the Association of American Geographers*, 101 (1), 107–130.

McCann, E. and Ward, K. (Eds.). (2011). *Mobile urbanism: Cities and policymaking in the global age*. University of Minnesota Press, Minneapolis.

Molotch, H. and Ponzini, D. (Eds.). (2019). *The new Arab urban: Gulf cities of wealth, ambition, and distress*. NYU Press, New York.

Morton, M.Q. (2011). The Abu Dhabi oil discoveries. [Online] Available: https://www.geoexpro.com/articles/2011/03/the-abu-dhabi-oil-discoveries [Accessed 10 November 2020].

Peck, J. (2011). Geographies of policy: From transfer-diffusion to mobility-mutation. *Progress in Human Geography*, 35 (6), 773–797.

Ponzini, D. (2011). Large scale development projects and star architecture in the absence of democratic politics: The case of Abu Dhabi, UAE. *Cities*, 28 (3), 251–259.

Ponzini, D., Ruoppila, S. and Jones, Z. M. (2020). What difference does democratic local governance make? Guggenheim museum initiatives in Abu Dhabi and Helsinki. *Environment and Planning C: Politics and Space*, 38 (2), 347–365.

Rapoport, E. (2014). Utopian visions and real estate dreams: The eco-city past, present and future. *Geography Compass*, 8 (2), 137–149.

Rapoport, E. (2015). Globalising sustainable urbanism: The role of international master-planners. *Area*, 47 (2), 110–115.

Rapoport, E. and Hult, A. (2017). The travelling business of sustainable urbanism: International consultants as norm-setters. *Environment and Planning A: Economy and Space*, 49 (8), 1779–1796.

Shwayri, S. T. (2013). A model Korean ubiquitous eco-city? The politics of making Songdo. *Journal of Urban Technology*, 20 (1), 39–55.

South China Morning Post (2018). Oil-rich Abu Dhabi's Masdar City: green oasis or green ghost town? [Online] Available: https://www.scmp.com/week-asia/business/article/2133409/oil-rich-abu-dhabis-masdar-city-green-oasis-or-green-ghost-town [Accessed 10 November 2020].

Sovereign Wealth Fund Institute (2013). Sovereign wealth fund rankings. [Online]. Available: http://www.swfinstitute.org/fund-rankings/ [Accessed 10 November 2020].

Tatchell, J. (2010). *A diamond in the desert: Behind the scenes in the world's richest city*. Sceptre, London.

TFB (2020). The future build. [Online] Available: http://www.thefuturebuild.com/. [Accessed 15 July 2020].

The Guardian (2016). Masdar's zero-carbon dream could become world's first green ghost town. [Online] Available: https://www.theguardian.com/environment/2016/feb/16/masdars-zero-carbon-dream-could-become-worlds-first-green-ghost-town [Accessed 10 November 2020].

Weber, M. (1964). *The theory of social and economic organization*. Free Press, New York.

World Bank (2013). GDP per capita. [Online] Available: http://data.worldbank.org/indicator/NY.GDP.PCAP.CD [Accessed 10 November 2020].

World Future Energy Summit (2020). Visit. [Online] Available: https://www.worldfutureenergysummit.com/visit#/ [Accessed 10 November 2020].

Wu, F. (2012) China's eco-cities. *Geoforum*, 43 (2), 169–171.

Xie, L., Flynn, A., Tan-Mullins, M. and Cheshmehzangi, A. (2019). The making and remaking of ecological space in China: The political ecology of Chongming Eco-Island. *Political Geography*, 69, 89–102.

Xie, L., Mauch, C., Tan-Mullins, M. and Cheshmehzangi, A. (2020). Disappearing reeds on Chongming Island: An environmental microhistory of Chinese eco-development. *Environment and Planning E: Nature and Space*, 2514848620974375.

5

A SMART-CITY EXPERIMENT

The case of Hong Kong

Introduction

A cartographer seeking to map the smart-city phenomenon would have to travel the world. Smart-city initiatives are taking place in many different regions, touching most continents and avoiding a geographical polarization (Karvonen et al., 2018a). There is not a specific area of the planet where the development of *smart cities* is prominent, and the conventional distinction between Global North and Global South is not able to grasp the boundaries of the phenomenon. Smart interventions are common in European and North American cities, just like they are in Chinese and Indian urban settlements. *Smart* is an adjective, seemingly universal, which policy-makers are attaching as a label to cities from Eurasia, Africa and the Americas, forming a complex puzzle still in the making. In this context, places far away from each other and characterized by remarkably unlike geographies, histories and cultures, are equated by the word *smart* which is emerging as one of the most powerful equalizers of the current urban age. *Smart* appears to drive the regeneration of existing urban spaces. Smart urbanism pervades centuries-old built environments located in Spain, Italy, UK, South Africa, Kenya, Sweden, Brazil, Indonesia, Austria, Chile, Australia and the US (Caprotti and Cowley, 2019; Cowley et al., 2018; Crivello, 2015; Dowling et al., 2018; Fernandez-Anez et al., 2018; Garau and Pavan, 2018; Grossi and Pianezzi, 2017; Jirón et al., 2020; Machado Junior et al., 2018; March and Ribera-Fumaz, 2018; McLean et al., 2016; Odendaal, 2016; Offenhuber, 2019; Parks, 2019; Ranchod, 2020; Tironi and Valderrama, 2018; Wiig, 2018). *Smart* can also be found in the genesis of new master-planned settlements. In this category, numerous cities built from scratch in China, India, Japan, Kenya, Morocco and the Philippines, bear the *smart-city* logo (Côté-Roy and Moser, 2019; Das, 2020; Dameri et al., 2019; Datta, 2018; Mouton, 2020; Trencher and Karvonen, 2019). Finally, *smart* permeates various scales of urbanization, materializing in megacities as well as in small towns and villages (Chambers and Evans, 2020; De Falco et al., 2019; Höffken and Limmer, 2019; Spicer et al., 2019).

If a cartographer struggled to depict the complex spatial distribution of smart urbanism across the world, it would be even harder for a painter to illustrate the face of the smart city. This is due to the fact that smart urbanism does not have one face, but many. On the one hand, Information and Communication Technology (ICT), which is commonly regarded as the main ingredient of the formula for making a city *smart*, is a very diverse strand of technological development. ICT includes a plethora of technologies differing in shape, size, composition and scope, which range from micro sensors to mobile apps and from smart grids to pure software (see also Kitchin, 2016). On the other hand, the city is the most hetero-geneous typology of social organization that has ever existed. There is a myriad of cities on the planet, and not a single twin. Therefore, when the *smart* and the *city* collide, the result differs geographically from case to case. In addition, both ICT and the city are constantly evolving. The former as part of processes of technolo-gical modernization, and the latter in relation to multiple political, economic, social and environmental changes. Because of such inherent transformations, the practise of smart urbanism is like a kaleidoscope: it keeps changing forms and colours, enchanting and confusing the observer at the same time.

The promises made by contemporary smart-city initiatives are tempting. Vienna, for example, seems to approach smart urbanism 'systematically' and 'through comprehensive innovation,' in order to provide 'the best quality of life for all inhabitants' (Smart City Wien, 2019: no page). Barcelona, as a smart city, appar-ently 'improves quality of life for its citizens across the whole society' (Barcelona City Council, 2017: no page). The smart-city agenda of Milano, so the official narrative goes, 'leads to a better quality of life through effective, accessible and intelligent tools aimed at the optimisation of resources for all citizens' (Milano Smart City, 2017: no page). However, critical scholars working in the field of smart urbanism have been aware of the kaleidoscopic nature of smart interventions, and of the perils that can lie behind the promise of being *smart*. Empirical studies have shown that the implementation of smart-city projects is often chaotic, unsus-tainable and fundamentally disloyal to the claims of developers and stakeholders (see, for instance, Coletta et al., 2019; Cugurullo, 2018; Guma and Monstadt, 2020). Entering inside presumed smart cities and exploring empirically what is actually happening on the ground is therefore essential (Karvonen et al., 2018b; Kitchin, 2015; Shelton et al., 2015; Wiig and Wyly, 2016). In reality, smart urbanism might be happening accidentally. It could be the messy sum of existing and originally disconnected urban initiatives which are 'corralled into the semblance of an over-arching, coordinated, and branded narrative,' rather than a rigorously planned experiment (Coletta et al., 2019: 350; Dourish, 2016). It might also not happen at all. Smart urbanism could be a masquerade orchestrated to hide the agendas of the usual suspects, or to assist the entrance of new forces onto the stage. A disguise to cover up power struggles among politicians or the business plans of private com-panies. Only time and, above all, place will tell.

This chapter resonates with the above concerns, and uses the smart-city agenda of Hong Kong as a case study to investigate in empirical terms an actual practise of smart urbanism. Hong Kong is presented as a case study for three main reasons. First, it hosts a large-scale programme of smart urban development. In Hong Kong, smart urbanism occurs in flats, buildings and districts belonging to diverse urban areas, which means that within the geographical context traversed by the chapter *smart* can be observed in a variety of spaces and shapes. *Smart* is also physically manifested in Hong Kong by means of a plethora of different smart technologies. In other words, because of its broad scale and scope, Hong Kong's smart-city agenda, when empirically unpacked, reveals many facets of smart urbanism. Second, the smart interventions that take place in Hong Kong, have an impact which goes over the regional scale. The magnitude of Hong Kong's smart urbanism cuts across and beyond mainland China, becoming the concern of other countries in East Asia. In this sense, the case study encapsulates and explains the irregular and often hidden geographies of the smart-city phenomenon. Third, the experience of the government of Hong Kong in smart urbanism is relatively old. The initial smart-city agenda was launched in 2013 and then revamped in 2018. Smart urbanism is a highly mutable and 'long-term endeavour' and, therefore, its assessment should be conducted not simply in a specific place, but also through time (Karvonen et al., 2018c: 291). Having existed for several years, the smart-city story of Hong Kong offers this opportunity.

The rest of the chapter is structured as follows. The next section discusses the context where Hong Kong's smart interventions are implemented. The emphasis is on the neoliberal nature of the political economy of the region, and on the shadows that neoliberalism casts on the growth and sustainability of the city. The exploration of the local context forms the vantage point from which the emergence of a regional smart-city agenda is observed. Here the chapter illustrates the evolution of smart urbanism, in relation to the changing economic priorities of the government. It reveals how policy-makers target the development of different smart technologies, as a strategy to galvanize specific economic sectors such as ICT, artificial intelligence and biotechnology. Subsequently, the chapter explores the planning process through which the smart-city programme of Hong Kong is put into practise. There is actually a profound discrepancy between the grandiose claims of the government and the reality of smart urbanism. The book exposes the uncoordinated approach that local planners have towards the smart city, showing that, from a geographical perspective, *smart* covers only small and disconnected segments of the region. In the penultimate section, the lens of the analysis focuses on a single smart intervention called the Hong Kong Science and Technology Park where automated building management systems, artificial intelligences and anti-ageing biotechnologies are developed, in response to the everchanging economic goals of the government. The chapter ends with a reflection on the sustainability challenges that the Hong-Kong-style smart urbanism poses. *Smart* is not the last step that cities will take, and its evolution into artificially intelligent urban technologies generates new problems while ignoring crucial social and environmental issues which have existed for a long time.

The context and lineage of smart urbanism in Hong Kong

In 1841, Lord Palmerston, the then British foreign secretary, described Hong Kong as a 'barren rock with hardly a house upon it': a remote corner of the planet barely touched by urbanization (Shelton et al., 2013: 2). During the same year, Britain had occupied Hong Kong Island, and Lord Palmerston did not see much business potential in the new Eastern possession of his queen. He did not see much growth potential at all. Lord Palmerston, however, was wrong. That empty rock was going to become one of the most powerful financial and trade centres in the world, and a dense urban region with more skyscrapers than New York (Shen, 2008; The Economist Intelligence Unit, 2019). In the space of a century, Hong Kong experienced a fast and intense process of urban development, in sync with a series of land-reclamation initiatives which have exploited the urban potential of what is a largely mountainous territory. If Lord Palmerston could look at the skyline of today's Hong Kong, he would see an impenetrable wall of high-rise buildings, and very little would appear familiar. The urban geography of the Special Administrative Region (SAR) has changed deeply, and so its political structure and life.

Today, Hong Kong is a SAR of the People's Republic of China which it rejoined in 1997 without losing its pre-unification politico-economic system: capitalism. Labelled a *flawed democracy* by The Economist Intelligence Unit (2018), it is officially governed by the Hong Kong Government which is led by the Chief Executive (the highest political actor in the SAR), under increasing tensions with the Chinese Communist Party. As the current Chief Executive, Carrie Lam, put it in a recent talk, she 'has to serve two masters by constitution, that is the Central People's Government and the people of Hong Kong,' within a dynamic political landscape in which it is unclear where the influence of the Chinese Communist Party exactly starts and ends (Reuters, 2019: no page). What is clear and special about Hong Kong as a SAR, is that it largely operates under the principle of *One Country, Two Systems*. This means that although the city-region is politically part of China, from an economic perspective Hong Kong is not subjected to mainland China's socialism. Described by scholars such as Cheung (2000) as *laissez-faire capitalism*, the economy of Hong Kong has few governmental restrictions. According to the Index of Economic Freedom (2020), Hong Kong has the world's second freest economy. The intervention of the government in the economic activities that take place across the SAR is minimum, and great freedom is given to the many forces constituting Hong Kong's private sector. This *laissez-faire* attitude strongly impacts on finance and real estate, which form the bulk of the regional economy and the engine of urbanization (Haila, 2000; Lai, 2012; Shen, 2008). When it comes to urban development, in particular, the same politico-economic logic opposing governmental regulation and interference applies. As Raco and Street (2012) point out, the planning system of Hong Kong is flexible, driven by private investments and, in essence, very similar to that of a Western neoliberal city like London. The SAR owns the territory, and developers bid for parcels of land in the absence of a governmental vision of the city (see also Raco and Gilliam, 2012). Once a bid is won and a lease is signed, the government has little or no authority over the design and function of what will be built on a parcel: private developers have the first and the last word.

Over the years, the outcome of this uncoordinated model of urban development has been a heterogeneous and fragmented built environment. In terms of urban design and architecture, Hong Kong is a melting pot of different aesthetics. Modernist skyscrapers can be seen next to old Taoist temples, and run-down social housing estates stand in the shadow of colossal five-star hotels (Mathews, 2011). While such sheer architectural diversity, in contrast with the green of the peaks and the placid water of the harbour, might please the eye of the observer, the situation is dire from a housing perspective. Seeing Hong Kong and living Hong Kong are incomparable experiences. Academics and journalists have repeatedly stressed that 'the construction industry in Hong Kong has long been associated with poor quality,' particularly in relation to homes (Chiang and Tang, 2003; LSE Cities, 2011; Tam et al., 2000: 437; The Guardian 2013, 2017). Because of the regional chronic lack of developable flat land, Hong Kong's developers tend to maximize the space at their disposal by building compact high-rise housing units. Inside these buildings, are numerous extremely small flats. Official statistics indicate that in Hong Kong, the average living space per person is 13.2 m^2 (Hong Kong Government, 2017). In addition, in order to reduce construction costs and save money, developers often employ low-quality building materials and infrastructures (Chan et al., 2002). In the poorest areas of the SAR, it is possible to find 46 m^2 deteriorated apartments where several families and strangers live together. Dangerous and unhealthy, these precarious accommodations can host up to 30 residents. They are called *coffin homes*.

Housing is not the only pressing urban issue which Hong Kong is facing. The sustainability of the SAR is compromised by several interconnected problems which have the urban as the common denominator. In terms of environmental sustainability, for instance, Hong Kong is experiencing high levels of air pollution. The maze-like built environment of Hong Kong creates a giant urban dome which hinders air circulation and traps pollutants (Higgins, 2013; World Bank, 2015). Hong Kong's water, sea water in particular, is also severely polluted. The city-region has one of the busiest container ports in the world, which represents a constant source of traffic, waste and, ultimately, a major threat to both human health and biodiversity (Wong and Wan, 2009; WWF, 2015). Socially, Hong Kong is characterized by an increasing Gini coefficient which is symptomatic of a deeply divided city where the distribution of environmental burdens and benefits is unequal (Hong Kong Government, 2012). The SAR's housing market is the most expensive in the world, and over 200,000 people are forced to live in subdivided flats, with 65 per cent of families staying in units ranging from 7 to 13 m^2 (Hong Kong Government, 2016). In practise, only a minority of high-income workers can afford adequate housing and a clean and safe accommodation.

To date, the attempts of the government to improve the urban sustainability of the SAR have been futile. Emblematic is the case of *Hong Kong 2030*, the first large-scale programme of sustainable urban development in the history of the SAR, launched by the local planning department in 2007 under grandiose expectations that were never going to be fulfilled (Cugurullo, 2017). Among its key objectives were 'strengthening' Hong Kong's 'role as a global and regional financial and business centre' while, at the same time, 'conserving the natural landscape, preserving cultural heritage, meeting housing and community needs' and 'promoting arts, culture and tourism' (Hong Kong Planning Department, 2007: 22, 23). However, as noted by several scholars, the initial steps of the SAR towards a sustainability agenda ended up as a sequence of empty discourses promoting economic growth in synergy with social justice and environmental preservation, without any real political power animating them (Francesch-Huidobro, 2012; Higgins, 2013; Wong and Wan, 2009). In the end, *Hong Kong 2030* was not compulsory and the neoliberal game of private developers did not change, as no regulation of the built environment was imposed by the government. Likewise, the fragmented model of urban development responsible for much of the unsustainability of Hong Kong was left unaltered.

It is in this context that in September 2013 the government of Hong Kong released a smart-city agenda. The programme, called *Smarter Hong Kong, Smarter Living*, was meant to function as the leading framework for urban policy in the region. Ambitious in terms of scale and scope, it was described by the Secretary for Commerce and Economic Development as a 'blueprint': a detailed plan of action supposed to guide the creation of new urban spaces and the regeneration of existing ones, homogeneously across the whole SAR (Commerce and Economic Development Bureau, 2013: 1). Its key ideas reflected now common themes in smart-city discourses, and were based on an adamant faith in Information and Communication Technology as a medium to better understand and ultimately improve the metabolism of the city. More specifically, according to an influential policy document consulted during the fieldwork, the ideal smart city targeted by *Smarter Hong Kong, Smarter Living* was 'one with wide application of new technologies such as sensors, Internet of Things, cloud computing, and big data analytics' employed to increase 'efficiency in the use of resources' in transport, safety, healthcare, governance and economics. This document, informally called by local policy-makers the *Smart City Report*, was until 2018 the most authoritative reference on smart urbanism produced by the government. Apart from laying out, in broad strokes, the SAR's approach to smart urbanism, it established a conceptual link between the development of smart-city initiatives and the development of the country where they take place, stating that 'the smart city is a development opportunity' which should be used 'to contribute to Hong Kong's overall level of development.' It was circulated from the top offices of the SAR, down to a plethora of departments (including the Planning Department), to set the official tone of smart urbanism in Hong Kong. In essence, the document defined the meaning of the term *smart city* and identified the role of the smart city within the development of Hong Kong, expressing the will of the government to use *smart* as a piece to complete a much bigger puzzle.

While the smart city was not defined in a specific and paradigmatic way, one aspect was immediately carved in stone. Since its genesis, *Smarter Hong Kong, Smarter*

Living was explicitly connected to a broader agenda of regional economic development. In the same policy document explaining how smart urbanism is interpreted by the Hong Kong Government, ICT is presented not simply as a medium to realize a smart city, but also as an emerging economic sector which the government wants to cultivate. As stated in the document,

> Information and communication technology is not only a key enabler underpinning Hong Kong's thriving economy, it is also taking shape as an economic sector in its own right. The "Smarter Hong Kong, Smarter Living" strategy sets out the framework for Hong Kong to leverage new technologies to propel continuous economic development.

In 2015, these words were followed by policy action when the Hong Kong Government released a new policy agenda in which it authorized the investment of US$ 5 billion in the ICT sector, in order to diversify and foster the economic development of the SAR. This policy agenda was produced by the Central Policy Unit, one the highest offices in Hong Kong, whose responsibility is to advise the Chief Executive on every matter concerning the development of the SAR, and to coordinate all the policies implemented in the region. The Unit was revamped in 2018, and it is now called the Policy Innovation and Co-ordination Office. It is responsible for the emergence of the *smart city* concept in the development agenda of Hong Kong. Every public initiative related to smart urbanism in the SAR depends on its agency. When one of the policy-makers behind *Smarter Hong Kong, Smarter Living* was asked in an interview about the origin of the SAR's smart-city initiatives, her reply was: 'The direction comes from the Central Policy Unit. They started everything and they are leading everything.'

In 2018, smart urbanism in the SAR evolved into *Hong Kong Smart City Blueprint* which is the latest incarnation of the regional smart-city agenda (Hong Kong Government, 2018a). This new programme takes from its predecessor the ambition of forging a blueprint, and advances a master plan to turn the entire region into a smart city. The overarching vision continues to be grounded in ICT as a silver bullet to address urban challenges. However, what differs is that in *Hong Kong Smart City Blueprint* the concept of *smart* goes beyond just ICT, and the boundaries of smart urbanism are pushed to include emerging technologies. In addition to traditional smart technologies, such as sensors and smart grids, the new master plan features artificial intelligence (AI), robotics and nanotechnology, thereby opening up novel and more experimental strands of smart urbanism. The 2018 smart-city agenda of Hong Kong, compared to the 2013 incarnation, manifests a rapid innovation in terms of technological contents. In five years, the SAR's smart urbanism has been updated to incorporate and reflect the most recent innovations not only in ICT, but also in cognate fields.

Furthermore, this new smart-city agenda is a reflection of the new political economy of Hong Kong, which has also been updated. If in 2013 ICT was in the SAR an emerging economic sector requiring government investment, nowadays this is not the case anymore. After having spent more than five years nurturing the ICT sector with a series of investments (such as the already mentioned US$ 5 billion to roll out *Smarter Hong Kong, Smarter Living*), the government of Hong Kong is now looking for new areas of economic expansion: economic sectors to focus the next wave of government investments. Artificial intelligence is one of these areas. In 2018, the Hong Kong Government (2018b) recognized that while AI research and technology had been developing rapidly around the world, Hong Kong was lagging behind, particularly compared to mainland China and other Asian countries like Singapore and South Korea. This recognition echoed the warning of local AI experts and journalists, stressing in chorus that 'Hong Kong must innovate or die' (SCMP, 2018: no page). Such concerns about the course of the economic development of Hong Kong, were soon followed by *ad hoc* policies. The government budgeted $5 billion to be spent immediately in 2018, with the aim of supporting AI research and the growth of private companies specialized in AI technologies, through funding schemes and incentives (Hong Kong Government, 2018c). 'The Government has now set AI as one of the key areas of technological development and will devote more resources to enhance Hong Kong's research and application capabilities in the AI field' declared the Secretary for Innovation and Technology in October 2018 (Hong Kong Government, 2018c: no page).

These injections of government investment synchronized with the economic strategies of the highest offices of the SAR, are intrinsically connected to the rolling out of smart interventions in the territory. As this section has illustrated, in Hong Kong, the smart-city agenda mirrors the agenda for regional economic development. The Hong Kong Government's directions in terms of political economy have steered the evolution of smart urbanism, and the way in which *smart* has been interpreted and translated into policy has been recurrently influenced by economic rationales. So far, the chapter has unpacked the lineage of Hong Kong's smart urbanism in relation to the broader context of the SAR, shedding light on the neoliberal character of the city-region and the many sustainability challenges that undermine its present and future. This contextual perspective will be maintained in the following section which will delve into the materialization of the smart-city agenda in the built environment, revealing its spatial distribution. Over the years, Hong Kong's understanding and practise of smart urbanism have been filtered by a complex urban, political and economic geography. Through this process of contextual filtering, *smart* has changed, but it has also caused change by driving the genesis of new urban areas and the transformation of existing ones. The narrative will now explore the implementation of smart urbanism across the SAR, moving beyond ideals and overarching political economies, to identify what urban spaces and technologies have been actually produced by the Hong Kong Government's smart-city agendas.

The planning and geography of smart urbanism in Hong Kong

At the regional scale, there is a substantial discrepancy between what *Hong Kong Smart City Blueprint* promotes and the reality of smart urbanism in the SAR. No overarching blueprint for disciplining the implementation of smart urban spaces across the region actually exists. As a representative from the local planning department revealed in an interview, 'we don't have an integrated policy. Our approach is project-based.' She explained that the Planning Department has never produced a master plan, and it does not intend to work on a detailed plan of action for the development of smart technologies and infrastructures in Hong Kong. The focus of the Planning Department is on single independent projects which are not connected among each other, and are not related to a broader urban vision. In this sense, there is no well-defined representation (whether in a written or pictorial format) of how the Hong Kong Government imagines the ideal smart Hong Kong to be in terms of shape and contents. No planning document. No study. Not a single rendering. The reason behind such lacuna is that the government does not plan to develop a smart urbanism in the entire region. Different members of the Planning Department confirmed, during separate interviews, that the Hong Kong Government is rolling out its smart-city agenda only in two areas of the SAR. More precisely, as one of them stated, 'the focus is on the Central Business District Two and the Hong Kong Science and Technology Park.' These two urban areas can be observed in Figure 5.1, which shows how geographically limited is the scope of smart urbanism in Hong Kong.

The Central Business District Two and the Hong Kong Science and Technology Park both lie upon land which is being redeveloped. The Central Business District Two (CBD2) is an area undergoing a significant process of urban regeneration. Most of the existing buildings are supposed to be torn down in order to be replaced by brand new smart buildings and infrastructures. For this zone, the main objective of the Planning Department is the creation of office spaces. 'We are short of office supply' pointed out an interviewee, a policy-maker working on the regeneration of CBD2.

> We are thinking of taking the opportunity that Kowloon East offers with the Kai Tak development, which is the old airport, and also the old industry area in Kowloon Bay and Kwun Tong, that we can change into business use. We are clustering these areas together and naming it CBD2. We want to regenerate this area, to make it more appealing to business and attract companies and multinationals.

The Hong Kong Science and Technology Park, instead, takes place in a plot which was previously untouched by urbanization, and can be therefore used by the government as a blank canvas to test experimental types of smart urbanism. 'Here we want to encourage research and technological innovation particularly in the field of Information and Communication Technology' stated a member of the Planning Department. She specified that this initiative is in line with the directions established by the Central Policy Unit which the Planning Department is obliged to serve. The interviewee referred, in particular, to the policy agenda set by the Central Policy Unit in 2015. This is the year when, as explained earlier in the

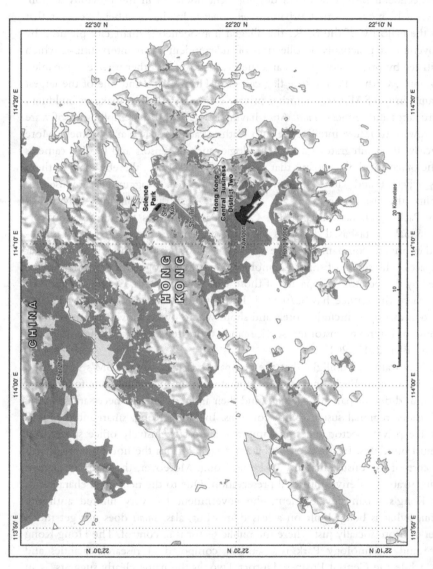

FIGURE 5.1 Map of the actual geography of smart urbanism in Hong Kong. The areas in black represent the only portion of the SAR, where the government is implementing its smart-city agenda.

Source: Author and the University of Manchester' Cartographic Unit.

chapter, the Hong Kong Government concentrated its investments on ICT, as well as the time when the interview occurred.

The remaining share of the urban territory of the SAR (highlighted in dark grey in Figure 5.1), has been extensively parcelled and leased over the years, according to the neoliberal model of urban development discussed in the previous section. Here what happens is smart urbanism *on demand*. In the neoliberal context that affects the majority of the SAR, the alleged homogeneous smart city pictured by the government is actually a collection of independent smart interventions which are initiated by private companies and then carried out by other private companies. For example, Cathay Pacific (the flag carrier of Hong Kong and one of the largest companies in the SAR) pays another private company, IBM, to make its buildings, infrastructures and services *smart*. After having being paid, IBM employs its software and hardware to infuse intelligence into Cathay Pacific's spaces and business. More specifically, IBM integrates smart technologies such as sensors and CCTV cameras into the offices of the company that is paying for its service, thereby making them energy efficient and safer. Similarly, IBM's purchased programmes are used to streamline the storage and analysis of the data that the economic activities of Cathay Pacific generate. Eventually, Cathay Pacific utilizes the information collected to make data-based decisions over its commercial operations. Likewise, several other private companies engage with IBM to bring smart urbanism into their own spaces. The scope of smart technologies and their geographical boundaries are dictated by the economic capacity of those who pay for them. *Smart* goes and acts solely where the service has been paid. In addition to the case of Cathay Pacific, other local examples include hotels and shops where owners invest in smart interventions to improve customer service and, ultimately, their revenue (see, for instance, IBM, 2017, 2018).

It is important to highlight that the companies that consume IBM's services do not act in concert. Each one of them operates on an individual basis for its own benefits, in different parts of the city, and their interests are not necessarily in sync with broader regional sustainability concerns. In Hong Kong, smart interventions led by the private sector affect and improve almost exclusively office spaces, not residential buildings. In this sense, they are detached from the housing crisis that is at the core of the unsustainability of Hong Kong. Moreover, they are not aligned with any strategy of environmental preservation. Due to the neoliberal character of Hong Kong's urban development, the government has very limited authority regarding what is being built on a leased area but, alas, *smart* does not get much greener or more socially just where the public sector is in control. The Hong Kong Science and Technology Park is essentially composed of research facilities and offices, while the Central Business District Two, as the name clearly suggests, is an economic space meant to accommodate the offices of multinationals. Hong Kong's public and private smart interventions have a common denominator: business.

None of the new smart districts developed by the Planning Council targets a social goal. From an environmental point of view, *smart* meets *sustainable* only in the so-called *smart building*: an architectural structure equipped with automated management systems and generators of renewable energy. According to a policy report on smart buildings disclosed during an interview with a senior planner from the Planning Department, 'the smart building is one of the marks of a smart city. Both the Government and the construction industry promote sustainable green buildings to make efficient use of energy and other resources and to mitigate the overall environmental impact.' The document illustrates the benefits that smart buildings can bring in terms of environmental sustainability, by reducing energy waste and carbon emissions. Subsequently, it states the intention of the Planning Department to apply this type of architecture to the development of the Central Business District Two and the Hong Kong Science and Technology Park. However, geographically, smart buildings decarbonize only a small percentage of the region because, as shown in Figure 5.1, this is an architecture which is present in a fraction of the SAR's territory. Above all, these are smart solutions tackling environmental issues of urban energy and CO_2 levels which, although important, do not include equally pressing problems of urban ecology and regional biodiversity. The policy report in question made no reference to ecological issues.

Behind the choice of the government to ignore crucial ecological concerns, in the same way that urgent social issues like housing are disregarded, is the intention to prioritize economic objectives. The imperatives set by the Central Policy Unit to direct the economic development of Hong Kong, steer the SAR's smart urbanism towards a type of urban planning that is almost indistinguishable from economics. Consequently, the *modus operandi* of the Planning Department manifests an acute predilection for economic matters. In this regard, emblematic are the aims and positions of the planners from the Planning Department interviewed during the fieldwork. 'We are trying to change the environment to make it more conducive to business' maintained a planner, while discussing the regeneration of the Central Business District Two. 'We need to make the environment attractive to companies and the smart-city concept seems to be the way forward' commented another planner, in relation to the development of the Hong Kong Science and Technology Park. 'Our mission is to drive technological innovation and support the economic growth of Hong Kong' summarized a chief planner. The objectives are economic in nature, and it is important to note that these are not abstract discourses: they are orders and inputs which shape the city, creating tangible structures that ultimately impact not only on the economy, but also on society and the natural environment. The next section shifts the scale of analysis, increasing the focus and moving the narrative to the specific case of the Hong Kong Science and Technology Park, in order to explore more in-depth the materiality of smart urbanism in the SAR.

The Hong Kong Science and Technology Park

The Hong Kong Science and Technology Park (HKSTP) is one of the flagship smart-city initiatives of the SAR, promoted by the government. It consists of a 220,000 m^2 built environment designed to host over 800 high-tech companies working on a wide range of smart technologies (see Figure 5.2). The official mission of the Park is to 'catalyse technological innovation' in a broad spectrum of fields, spanning from robotics to renewable energy, and from artificial intelligence to nanotechnology (HKSTP, 2018a: no page). From a geographical perspective, as Figure 5.1 shows, the Park is located in Pak Shek Kok in the North-East part of the region. It is detached from the main urban areas of Hong Kong, while being close to the campus of the Chinese University of Hong Kong with whom the Park has a research partnership. Its position has been strategically determined by a two-fold economic rationale. First, the area where the HKSTP is being developed is much cheaper compared with Kowloon (the hyper dense centre of the SAR) and the northern side of Hong Kong Island. On these terms, Pak Shek Kok, which is relatively remote hence less expensive, represents a major source of savings for the developers. Second, the Park is close to mainland China. There is a direct connection linking Shenzhen (an important financial node adjacent to the southern border of China) with the HKSTP, via the regional metro system. Moreover, Shenzhen and the Park are connected by a major road passing between two ridges, which allows for a quick transit by car.

As a manager from the HKSTP explained during an interview, the geographical configuration of the Park (especially its proximity to mainland China) is convenient and profitable because Chinese investors and potential customers can easily reach it and see its showrooms. In essence, the Park is a space where private companies develop and test experimental smart technologies later showcased inside and outside

FIGURE 5.2 The heart of the Hong Kong Science and Technology Park.
Source: Lee Yiu Tung

the many buildings that form it. The HKSP features state-of-the-art architecture which integrates smart sensors, hybrid energy generators combining solar with wind power, smart grids, photovoltaic panels and, above all, a central software regulating automatically lighting, heating and ventilation systems. On its streets, pilot models of self-driving cars are tested. Promotional events are organized on a daily basis, and the doors of the Park are always open for those interested in buying the technology that is there developed, or investing in the research projects that are there conducted.

The story of the Science Park exemplifies Hong Kong's uncoordinated approach to smart urbanism, as well as the interlinkages between the SAR's smart-city agenda and the government's agenda for economic development, discussed in the previous sections. One of the directors of the HKSTP clarified in an interview that 'the Park wasn't created with anything like a smart-city agenda in mind,' stating that its construction was not part of any regional programme of urban development, but rather a stand-alone initiative. This statement is confirmed by the fact that the Park was already operational in 2005: eight years before the release of *Smarter Hong Kong, Smarter Living*. He continued saying that 'it does not follow any urban policy in particular,' and regulates itself. Technically, the HKSTP is directed by a public company called Hong Kong Science and Technology Parks Corporation, which is also responsible for the management of three industrial estates owned by the government and another science park co-developed with mainland China (the Hong Kong-Shenzhen Innovation and Technology Park). The main roles of this public corporation are to devise the agenda of the Park and to manage its facilities. More specifically, the board of directors of the corporation, appointed by the Chief Executive and the Financial Secretary (the Minister of Finance), decide what research themes the Park will concentrate on. This decision, one member of the board explained, is always based upon economic directions defined by the Central Policy Unit. As he put it in an interview, 'the Park was created to stimulate a more diverse economy, and now we are here to help diversify the economy and help boost employment in new sectors like ICT and green technology.' This interview was conducted in early 2016, and the director made explicit references to the 2015 policy agenda set by the Central Policy Unit to cultivate the ICT sector of the SAR. The research themes chosen by the board recurrently target gaps in the economy of Hong Kong, where the government sees potential. Once these themes are identified, the Park welcome both emerging and established private companies whose agenda is in line with them.

The Science Park's research themes have been changing, and their mutable nature reflects changes in the political economy of Hong Kong. Up to 2017, they were *ICT, precision engineering, electronics* and *green technology*. In the words of a manager of the HKSTP at that time the shared goal was 'building automation.' During an interview that took place in the summer of 2016, he noted:

> Most of the companies in the Park are working on developing automated systems so that your light switches off when you leave the room, or you make sure the air con is on by the time you get home and the curtains are open.

This research agenda resonates with common practises and understandings of smart urbanism and, more particularly, with the notion of the *smart building*. It also resonates with the sustainability goals identified by the local planning department, according to which *smart* meets *sustainable* in the automated building. As seen in the previous section, the Planning Department was aiming to implement this architecture in the Park. The board of directors welcomed it, not as an architecture per se, but as a research strand in sync with the then economic directions of the Central Policy Unit. Essentially, the HKSTP was focusing on the development of building management systems capable of controlling, in an automated manner, key aspects of a building such as heating, ventilation, air conditioning, lighting, security and fire safety. In this field of research, smart sensors play a key role. In the Park's offices, for example, lights are automatically turned on when an occupancy sensor detects motion in a space. Similarly, the HKSTP's windows are automated and open depending on what temperature is perceived by indoor sensors, or when smoke is detected.

This research trend came to a halt in 2018 when two new themes began to reshape the Park's technological portfolio: artificial intelligence and biotechnology. The first theme reflects the evolution of the SAR's smart-city agenda, from its 2013 incarnation (*Smarter Hong Kong, Smarter Living*) to the 2018 manifestation (*Hong Kong Smart City Blueprint*). It signs the passage from ICT and now increasingly common smart sensors to emerging AI technologies. On these terms, the appearance of the AI theme is symptomatic of the technological update of the Park which has started a transition to the most recent hardware and software available on the market. In addition, it is a symptom of one of the latest updates of the political economy of Hong Kong. As noted in the second section, in 2018 the Hong Kong Government decided to invest in artificial intelligence. In practise, during that year, the HKSTP received US$ 2 billion from the government to boost research on artificial intelligence and support the growth of private companies working on AI technology (Hong Kong Government, 2018c). On top of that, the Park has managed to attract investment from mainland China by catching the attention of Alibaba, the Chinese tech giant specialized in e-commerce and, more recently, in AI products. Alibaba is by far the biggest company in partnership with the HKSTP. Since 2018, it has invested over US$ 227 million in Hong Kong's Science Park to build a research lab studying autonomous driving systems for commercial use. The Park provides the laboratories where Alibaba's researchers develop the software that animates self-driving cars, as well as the space where pilot models are tested. Beyond autonomous vehicles, Alibaba is also using the HKSTP to experiment with more extreme artificial intelligences. Its latest innovation in the field of AI is called City Brain which is an autonomous traffic management programme (Alibaba, 2019). It draws upon data collected via hundreds of cameras and examined through real-time video analysis, to control the traffic lights of a given city (Cugurullo, 2020). Everything happens in complete autonomy.

The second theme, instead, has set a different direction of research, which does not fit with the shared goal that was identified by the HKSTP manager in 2016.

There is a considerable difference between building biotechnology and building automation. Companies like Veridian Biotechnology Limited and Pure Innovation Biotech Limited, for instance, have installed in the Park several laboratories specialized in the development of protein drugs (HKSTP, 2018b, 2018c). However, there is a clear compatibility between building biotechnologies and the recent will of the Hong Kong Government (2019: no page) to turn the SAR into a 'biotech centre.' Since 2018, the Science Park has received from the government over US$ 2 billion to nurture biotech companies working in the field of protein production. This is a rapidly evolving field of research dealing with the engineering of proteins, to treat a plethora of medical conditions and overcome traditional human limits such as ageing. For example, a type of protein that is being researched in the Park is called Sirtuin 1 (SIRT1) which is colloquially referred to as the *anti-ageing protein*, because it delays the ageing of cells. Over the course of life, the quantity of Sirtuin 1 in the human body naturally decreases. Biotechnology, in the shape of protein drugs, can stimulate the SIRT1 in a given body to boost its anti-ageing capacity, as well as create and insert new SIRT1 proteins. It is not surprising that the prospect of delaying the physical decay of the body and challenging death makes this technology highly in demand. Likewise, it is not a surprise that the Science Park has managed to attract investment from China to cultivate this strand of research and commercialize its outcomes internationally.

Overall, Hong Kong's Science Park is a microcosm where the deepest tensions inside the SAR's smart urbanism come to light, particularly in terms of sustainability. The Park is a space built to realize the economy envisioned by the government. In this sense, it is a vehicle meant to reach economic targets, and its technological asset is dynamic because the Hong Kong Government keeps adjusting the trajectory of the SAR's economic development. When the Central Policy Unit pushed for the development of the SAR's economy towards ICT, the Park pushed for the development of information and communication technologies. Later, when the government targeted biotech, the Park experimented with biotechnologies. From ICT to AI and from automated buildings to protein drugs, the HKSTP's portfolio has included new items matching the politico-economic decisions taken by the SAR's highest offices. The board of directors set research themes whose twofold purpose is to attract private companies, and to facilitate their clustering into a research area which is the mirror image of the economic area where the government wants to invest. *ICT* and *Biotech*, for example, are clusters of the HKSTP, as well as economic sectors of the SAR. 'Whenever a partner company moves in, we assign it to a cluster,' explained a member of the board, 'so they can network with other companies in the same field.' The goal is not the growth of a single company, but rather the communion of different companies and the cultivation of a sector of the regional economy, pre-identified by the Central Policy Unit.

The stark economic inclination of the Hong Kong Science and Technology Park flourished at the expense of its environmental and social dimensions. The implementation of the Park, for instance, followed what a representative from the HKSTP called 'the standard approach in Hong Kong.' According to the interviewee, the building process was not informed by any ecological study. The developers did not conduct an environmental impact assessment, and simply cleared the plot of the existing vegetation. They then levelled and paved the land, ultimately creating a large expanse of concrete and steel, insensitive to issues of biodiversity loss and ecosystem services. Moreover, when observed from a social angle, the Park is essentially an isolated urban environment populated almost exclusively by the people who work there, with little or no social activities after office hours. It is composed of offices and laboratories devoted to monetizing technological innovation. Its social dimension is thin, and the contribution that it offers to the problems affecting the social sustainability of Hong Kong (the housing crisis, in particular) is nonexistent.

The detachment of the HKSTP from the rest of Hong Kong is an issue that cannot be explained only in relation to the remote geographical location of the Park. It is a problem also going beyond the type of spaces that the SAR's smart-city agenda offers. Even if the Science Park was located in the centre of Hong Kong and provided, in addition to offices and labs, numerous accommodations, its core activities would still not contribute to the sustainability of the region. The problem lies not in the geography of the Park, but in the geography of the technology that the Park produces. For over a decade, the private companies based in the HKSTP have created products which have been integrated exclusively into a minority of buildings, thus disregarding the city as a whole. HKSTP-branded technologies have often not even targeted an entire building, limiting their effect to just a single flat over many. The crux of the matter is that *smart* is intervening in a fraction of the city, with no coordination and without feeding into a broader process of urban development. When, during an interview, an HKSTP manager was asked to reflect on how disconnected the technologies produced in the Park are, his reply was:

> That's absolutely right. The larger focus has been on single technologies and single buildings. There is a lot of promotion now of home products so that you can use your smartphone to switch on the aircon before you get home. And you are right, they are all quite limited footprints or spheres of influence.

He paused for a moment and then concluded: 'There is no understanding of the big picture. There is no understanding of the city as an organic entity.'

Conclusions

The exploration of the smart-city agenda of Hong Kong has shown that talking about smart urbanism by means of singular nouns is inappropriate. In the SAR, there is not such a thing as a smart *city*. The stress on the word 'city' is here meant to emphasize that a cohesive singular entity has never been produced by the smart-city programme of the government due to the fact that local policy-makers have never produced a cohesive smart-city programme in the first place. From an ontological perspective, this chapter has provided evidence of the fragmentation, heterogeneity and multiformity lying at the very foundation of an alleged smart city, thereby challenging the understanding of smart cities as organic urban experiments. Far from being a homogeneous whole, Hong Kong's smart urbanism is made of incongruous elements lacking a regular and methodological arrangement, and symptomatic of a *laissez-faire* approach to city-making. More specifically, *Smarter Hong Kong, Smarter Living* and *Hong Kong Smart City Blueprint*, are not systematic large-scale master plans of urban development, as the government's press releases tell. In reality, they are a fragmented patchwork of different and disconnected initiatives, carried out in different spaces, by different actors and for different motives. The aura of singularity or, in other words, the quality of being singular that surrounds the smart-city image crafted and officially promoted by the Hong Kong Government is, in practise, a façade hiding a messy and unsustainable landscape.

The reasons behind such fragmented experience of smart urbanism are deeply contextual. Because of the neoliberal context which traditionally characterizes the urbanization of Hong Kong, the majority of the urban areas of the SAR are privatized. Therefore, the government has scarce control over what is being built in most of its territory. The same dynamics apply to smart urbanism. Smart interventions do not bypass their geographical context. Instead, smart-city initiatives are shaped by the geography of the place where they occur. As a result, when in the *laissez-faire* urban realm of Hong Kong, neoliberalism meets smart urbanism, the outcome is *smart on demand*. Smart-city projects are initiated by private companies and then implemented by other private companies, exclusively in private spaces, for the benefits of the single rather than of the whole. The only two areas of the region (not yet privatized), where the Hong Kong Government is actively intervening in the attempt to achieve a smart urbanism, do not show homogeneity and connectivity between each other and in relation to the rest of the SAR. The Hong Kong Central Business District Two is a zone which is being regenerated to increase the regional supply of offices and attract multinationals. The Hong Kong Science and Technology Park is a collection of laboratories, offices and testing facilities, where a diverse technological portfolio, ranging from ICT to robotics and from AI to biotechnology, is being developed and trialled. Behind the genesis of the CBD2 and the Science Park, there is not a detailed vision of the smart city, which captures the desired shape and contents of the built environment. The planning process which oversees their development is loose and, unlike what the government's discourses try to officialize, a blueprint connecting and disciplining the SAR's smart interventions ultimately does not exist.

Just as illusionists know that there is no real magic in the trick they perform, the government is, to some extent, aware that its smart-city agenda is not an organic plan of regional urban development. This is evident in the opinions of the very people that are crafting and implementing the SAR's smart-city initiatives. The illusion of the smart city as a homogenous urban project is apparent to several local policy-makers and planners who understand well how chaotic *Smarter Hong*

Kong, Smarter Living and *Hong Kong Smart City Blueprint* actually are. 'The smart city here is not holistic or comprehensive' pointed out a member of the Planning Department in an interview. 'We are doing many different things but they are fragmented' recognized a policy-maker, while reflecting upon the rolling out of *Smarter Hong Kong, Smarter Living*. For some of them the problem is a matter of governance. 'The governance is not easy' argued a senior planner responsible for the implementation of *Hong Kong Smart City Blueprint*, complaining that 'a lot of private interests are involved,' and it is therefore difficult for the local planning department to impose an overarching vision of the city, capable of unifying all the heterogenous agendas at play. On these terms, smart urbanism becomes necessarily fragmented, because the interests driving it are also fragmented, and the formation of a homogeneous smart city is hindered by the absence of a homogeneous vision of the smart city. The missing element is a shared urban prospect: the possibility of a future city, with the capacity to reconcile divergent interests.

The above interpretation resonates with the words of one of the directors of the Science Park who stated in an interview: 'In Hong Kong, the smart city push is a challenge because it means bringing together different departments, offices and bureaus, sometimes with conflicting agendas.' This statement puts emphasis on the number of actors involved in the practise of smart urbanism, as well as on the critical lack of a shared agenda: a void which leads to division, rather than to cohesion. Multiple players pursue diverse goals, and generate forces that, instead of merging into a synergistic action, push the built environment to expand and change in an individualistic manner. These forces diverge from each other and, thus, they separate. The resulting fragmentation is also made possible by the uneven power relationships underpinning the urban governance of the SAR. Different actors have disparate agendas and, some of them, are in the position to foster their own interests and, above all, to completely disregard those of the competitors. The case of the Central Policy Unit and the Environment Bureau, for instance, is emblematic. The former has an agenda based on economics, while the latter's agenda is driven by the principles of ecology. The Central Policy Unit seeks economic growth. The Environment Bureau is responsible for preserving the SAR's biodiversity. No attempt is made to find a common ground, and the consequence of this lacuna is a scenario of governance prone to conflict. 'The two sides have conflicting priorities' reflected a policy-maker working in one of the SAR's top offices, concluding: 'At the end of the day, however, the Central Policy Unit is more influential than the Environment Bureau, that's certainly true.' The Central Policy Unit silences the voice of the Environment Bureau whose ecological expertise eventually does not feed into the planning of Hong Kong's smart interventions.

Some of these governance problems can be traced back to an issue of labelling, which involves policy-makers simply *naming* smart cities, without first *making* smart cities. Not only the government never formulated a detailed and shared vision of the smart city, but it also added confusion and chaos by taking already existing urban initiatives and attaching the *smart city* label to them. This is the case of the Hong Kong Science and Technology Park, for example, now officially part of the government's smart-city agenda, while its origin predates by a decade the genesis of *Smarter Hong Kong, Smarter Living*. For one of the directors of the Park,

> the smart city is a term which is being used by the government to harness many things which have already been happening. The smart city is a convenient name to bring it all together. Until recently those initiatives had not been given the smart city label, but now they are coming together under this umbrella of the smart-city movement. Largely because of what is happening over the sea in cities like Barcelona, Vienna and Malmo, and of course in mainland China, and Hong Kong as a global city wants to emulate that.

The intuitions of the director correspond to the narrative of the *Smart City Report* discussed in the second section, where Barcelona is described as 'an outstanding performer in smart-city development in the EU,' In addition to Barcelona, to which an in-depth 15-page annexe is dedicated, the report examines several other European, American and Chinese smart-city initiatives representing 'worthy examples of state-of-the-art experiences, for Hong Kong.' Most importantly, the *Smart City Report* clearly states that it is the job of the Planning Department to 'systematically coordinate the development of existing urban projects,' including the Science Park, and steer them 'towards the formation of a smart city.' With these words, the Hong Kong Government is setting the SAR's planners with the almost impossible task of creating a smart city out of ten-year-old urban projects (internally regulated and driven by their own agendas), in the absence of an overarching master plan.

However, as this chapter has stressed, while the government of Hong Kong might not have a detailed vision of the smart city, it certainly has a clear purpose for smart urbanism. Smart interventions are strategically employed for a well-defined politico-economic rationale. There is no large-scale blueprint in terms of urban planning, urban design, architecture and engineering, but there is an ambitious blueprint in relation to the political economy of the SAR, and the smart city is part of it. The Chief Executive and the Central Policy Unit have identified the sectors of the regional economy that they want to cultivate. In this regard, a smart intervention such as the Hong Kong Science and Technology Park is a medium to galvanize ICT, biotechnology and artificial intelligence, not as strands of smart urbanism, but as emerging economic sectors. The government has also defined the production of state-of-the-art office spaces as an economic priority to attract large multinationals to Hong Kong. To this end, a smart intervention like the Hong Kong Central Business District Two is not a mere act of urban regeneration meant to redevelop derelict buildings, but a chance to produce modern offices where valuable economic activities will take place. Meanwhile, the government wants to keep the regional economy free from strict governmental interventions, because this is why many private companies choose to do business in the SAR. Therefore, the government's smart urbanism does not clash with the neoliberal urbanism that dominates the territory, and the lack of regulation concerning smart urban projects in private spaces is not a casual oversight, but part of an intentional strategy of urban governance.

In essence, when it comes to smart urbanism, the Hong Kong Government does have a plan, but it is one made with the instruments of economics. It is not a master plan or a blueprint designed with the tools of urban planning, urban design, geography and engineering. *Smarter Hong Kong, Smarter Living* and *Hong Kong Smart City Blueprint* are plans of economic development, tactically connected to the regional political economy. Such grand design, however, appears to be deeply flawed when the angle of observation is not purely politico-economic. From a sustainability perspective, in particular, the stark business-oriented connotation of the SAR's smart urbanism, produces a plethora of problems. First, from a social point of view, there is no attention to urgent questions of justice and equity. Smart interventions are scattered across the SAR unevenly, and their advantages are confined mostly within the private spaces of the elite that can afford them. Smart flats, buildings and districts are a luxury in an increasingly expensive housing market. They are the exception for the exceptionally rich: dots on the map, rather than brushstrokes covering the entire city. The form of a smart intervention is also often that of the office, instead of the residential building. Hong Kong's formula for smart urbanism is producing offices and no alternative to the housing crisis affecting the region. It ignores the SAR's lack of affordable healthy homes, while it ameliorates the condition of the few high-income workers who already have access to adequate housing.

Second, in terms of environmental sustainability, the fragmentation intrinsic to the smart-city agenda of Hong Kong causes severe ecological fractures. Smart buildings, infrastructures and technologies are developed in isolation from the natural environment, and their construction disregards ecosystems and ecosystem services. The smart building, for instance, alleged by the SAR's policy-makers and planners to be the paragon of urban sustainability, is mute and deaf to ecological concerns. The HKSTP's smart buildings are an intelligent piece of architecture, inasmuch as they can process complex problems regarding their energy performance. However, they have no understanding of the surrounding bioregion and, in fact, they have been built over local ecosystems rather than in sync with them. *Smart* meets *sustainable* where it is economically profitable. The automated urban technologies produced by the Science Park are on the agenda because they are commodities that private companies can sell worldwide, and are installed in the CBD2 since they furnish premium office spaces for multinationals. *Smart* does not meet *ecological* because this is a juncture scarcely monetizable. Hence the so-called 'standard approach' in the unecological implementation of the HKSTP whose land was cleared of the existing flora: a methodology which reflects a broader trend in urban development, responsible for major damages to the ecosystems of China (Long et al., 2014; Qiu et al., 2015; Wan et al., 2015). Cases of soil erosion, loss of forest cover and wetland decline, due to an ecologically uninformed urbanization targeting economic growth, have been widely reported in China since the early 1990s (He et al., 2014; Peng et al., 2015). In this sense, what the Hong Kong Government promotes as an innovative smart urbanism only replicates the same patterns of urban development that have threatened Chinese bioregions for three decades.

Third, there are hidden geographies which Hong Kong's smart-city agendas do not take into account, although their sustainability-related ramifications are considerable. On the one hand, the direct sphere of influence of smart interventions is too narrow. *Smart* takes place in single flats, buildings and districts. Local planners and policy-makers do not attempt to employ the potential benefits of smart urbanism across the entire city, and there is no intention to scale up *smart* from the micro to the macro. Paradoxically then, the government praises and promotes a *smart city*, while there is actually no connection between the two elements of the concept, the *smart* and the *city*. The former belongs to the latter only tangentially. On the other hand, the indirect sphere of influence of smart interventions is broad and unrecognized by those who actuate and endorse them. Economically, for instance, a share of the positive outcomes of smart technologies and smart spaces physically located in Hong Kong eventually escapes the SAR. When IBM uses its ICT to make the business operations of Cathay Pacific smarter, the profit that the company based in Hong Kong generates is not exactly Hong Kong's profit. Because of its multinational nature, Cathay Pacific is composed of stakeholders which do not even live in Hong Kong. Moreover, under the local low-tax regime, it is only a minimal part of the company's income that feeds into the economy of the SAR. Similarly, the technology developed in the Science Park does not all stay in Hong Kong. Automated building management systems branded HKSTP are sold all around the world, and affect built environments that are far away from China. Biotechnology studied in the Park alters the bodies of wealthy individuals who are passing by Hong Kong to get a shot of anti-ageing protein. They will live their longer than average life in the place where they are coming from. Hong Kong is just a stopover to get *smart*, not the final destination. Environmentally, the actual ecological impact of the SAR's smart-city projects follows a geography crossing the boundaries of Hong Kong. The many smart devices installed in some of the buildings of Hong Kong are made of materials which do not come from Hong Kong. From a supply chain point of view, the production of smart technologies depends upon the extraction of metallic ores such as cobalt, which the SAR does not provide. The sources are mines in Central Africa where the actual environmental impact occurs (Nkulu et al., 2018; van den Brink et al., 2020). *Smart* is the last link of a long and hidden chain of environmental degradation.

In conclusion, it is worth remarking that the dynamics of smart urbanism have changed and will keep changing. Ideas and practises of smart city-making do not follow an immutable dogma, manifesting instead swift mutations. This chapter has explored the implementation of Hong Kong's smart-city programme over a period of six years, emphasizing how the contents of the government's agendas for a smart urban development have evolved, mirroring both the evolution of the regional agenda for economic development and recent innovations in the field of ICT and beyond. The emergence of artificial intelligence in the practise of smart urbanism, in particular, arguably casts significant winds of change. The creation of a software, such as Alibaba's City Brain, for example, has the potential to transform contemporary smart interventions in a twofold way, by reformulating their agency and scale. Compared to the automated building management systems that have been part of the SAR's technological landscape for over a decade, City Brain is supposed to operate in an autonomous manner. While a standard intelligent building reacts by following pre-programmed patterns and executes pre-defined actions, a city brain should act without sticking to commands and options previously determined by its creators. If such difference will actually materialize, as the policy-makers behind *Hong Kong Smart City Blueprint* are hoping and pushing for, it would mark the passage from automation to autonomy in the management of the city. In addition, the scale of smart urbanism might also differ, since the footprint of the majority of the ICT devices installed in Hong Kong cover only single flats, buildings and, in their largest manifestation, a district. The plan of the developers of City Brain, in contrast, is to set the sphere of influence of this artificial intelligence to entire cities. Hypothetically then, the operation of a city brain could be the broad brushstroke that Hong Kong is lacking.

The growing influence of artificial intelligence on the development of cities signs the entrance of new forces and agents in the field of smart urbanism. This, however, does not mean that the context in which *smart* (whatever its incarnation is) takes place will change. The key sustainability problems of the smart-city initiatives occurring in Hong Kong, as the chapter has noticed, are deeply contextual and reside within the politico-economic structure of the SAR. In this sense, a variation in the degree of technological innovation of Hong Kong's smart interventions, will not considerably change the status quo and, thus, unsustainability will persist. Similarly, urban AIs will pose new questions, but long-standing urban questions will remain unanswered. The practise of smart urbanism in Hong Kong is unsustainable because it is based upon an understanding of the smart city, heavily skewed towards an elitist and ecologically insensitive economic logic. On these terms then, improving smart urbanism should not be about updating its technological contents, but rather rethinking its mission and functions, by incorporating crucial social and environmental concerns for the city as an organic whole. These concerns are now missing in the basic *forma mentis* of smart-city developers whose current mindset seems incapable of processing issues of justice, equity and ecology. At the end of an interview with a policy-maker in charge of the future direction of the smart-city agenda of Hong Kong, the tables turned and the interviewee asked the following question: 'Equity is a utopian concept. How do you assess what is equitable in a city?' An artificial intelligence cannot answer this question, at least for now. Smart-city initiatives should strive to reach a situation in which humans can answer fundamental socio-environmental questions and act upon them, before autonomous city brains will do it.

References

Alibaba (2019). ET City Brain. [Online] Available: https://www.alibabacloud.com/et/city [Accessed 10 November 2020].

Barcelona City Council (2017) BCN smart city. [Online] Available: http://smartcity.bcn.cat/en [Accessed 10 November 2020].

Caprotti, F. and Cowley, R. (2019). Varieties of smart urbanism in the UK: Discursive logics, the state and local urban context. *Transactions of the Institute of British Geographers*, 44 (3), 587–601.

Chambers, J. and Evans, J. (2020). Informal urbanism and the Internet of Things: Reliability, trust and the reconfiguration of infrastructure. *Urban Studies*, 0042098019890798.

Chan, E. H., Tang, B. S. and Wong, W. S. (2002). Density control and the quality of living space: a case study of private housing development in Hong Kong. *Habitat International*, 26 (2), 159–175.

Cheung, A. B. (2000). New interventionism in the making: Interpreting state interventions in Hong Kong after the change of sovereignty. *Journal of Contemporary China*, 9 (24), 291–308.

Chiang, Y. H. and Tang, B. S. (2003). 'Submarines don't leak, why do buildings?' Building quality, technological impediment and organization of the building industry in Hong Kong. *Habitat International*, 27 (1), 1–17.

Coletta, C., Heaphy, L. and Kitchin, R. (2019). From the accidental to articulated smart city: The creation and work of 'Smart Dublin.' *European urban and regional studies*, 26 (4), 349–364.

Commerce and Economic Development Bureau (2013). Smarter Hong Kong, Smarter Living. [Online] Available: http://www.digital21.gov.hk/eng/relatedDoc/download/2014D21S-booklet.pdf [Accessed 10 November 2020].

Côté-Roy, L. and Moser, S. (2019). 'Does Africa not deserve shiny new cities?' The power of seductive rhetoric around new cities in Africa. *Urban Studies*, 56 (12), 2391–2407.

Cowley, R., Joss, S. and Dayot, Y. (2018). The smart city and its publics: insights from across six UK cities. *Urban Research & Practice*, 11 (1), 53–77.

Crivello, S. (2015). Urban policy mobilities: the case of Turin as a smart city. *European Planning Studies*, 23 (5), 909–921.

Cugurullo, F. (2017). The story does not remain the same: multi-scalar perspectives on sustainable urban development in Asia and Hong Kong. In Caprotti, F. and Yu, L. (Eds.). *Sustainable cities in Asia*. Routledge, London.

Cugurullo, F. (2018). Exposing smart cities and eco-cities: Frankenstein urbanism and the sustainability challenges of the experimental city. *Environment and Planning A: Economy and Space*, 50 (1), 73–92.

Cugurullo, F. (2020). Urban artificial intelligence: From automation to autonomy in the smart city. *Front. Sustain. Cities*, 2, 38. doi:10.3389/frsc.2020.00038.

Dameri, R. P., Benevolo, C., Veglianti, E. and Li, Y. (2019). Understanding smart cities as a glocal strategy: A comparison between Italy and China. *Technological Forecasting and Social Change*, 142, 26–41.

Das, D. (2020). In pursuit of being smart? A critical analysis of India's smart cities endeavor. *Urban Geography*, 41 (1), 55–78.

Datta, A. (2018). The digital turn in postcolonial urbanism: Smart citizenship in the making of India's 100 smart cities. *Transactions of the Institute of British Geographers*, 43 (3), 405–419.

De Falco, S., Angelidou, M. and Addie, J. P. D. (2019). From the 'smart city' to the 'smart metropolis'? Building resilience in the urban periphery. *European Urban and Regional Studies*, 26 (2), 205–223.

Dourish, P. (2016). *The internet of urban things*. In Kitchin, R. and Perng, S.-Y. (Eds.). *Code and the city*. Routledge, London.

Dowling, R., McGuirk, P. and Maalsen, S. (2018). Realising Smart Cities: Partnerships and economic development in the emergence and practices of smart in Newcastle, Australia.

In Karvonen, A., Cugurullo, F. and Caprotti, F. (Eds.). *Inside smart cities: place, politics and urban innovation.* Routledge, London.

Fernandez-Anez, V., Fernández-Güell, J. M. and Giffinger, R. (2018). Smart City implementation and discourses: An integrated conceptual model. The case of Vienna. *Cities*, 78, 4–16.

Francesch-Huidobro, M. (2012). Institutional deficit and lack of legitimacy: the challenges of climate change governance in Hong Kong. *Environmental Politics*, 21 (5), 791–810.

Garau, C. and Pavan, V. (2018). Evaluating urban quality: indicators and assessment tools for smart sustainable cities. *Sustainability*, 10 (3), 575.

Grossi, G. and Pianezzi, D. (2017). Smart cities: Utopia or neoliberal ideology?. *Cities*, 69, 79–85.

Guma, P. K. and Monstadt, J. (2020). Smart city making? The spread of ICT-driven plans and infrastructures in Nairobi. *Urban Geography*, 10.1080/02723638.2020.1715050.

Haila, A. (2000). Real estate in global cities: Singapore and Hong Kong as property states. *Urban Studies*, 37 (12), 2241–2256.

He, C., Liu, Z., Tian, J. and Ma, Q. (2014). Urban expansion dynamics and natural habitat loss in China: a multiscale landscape perspective. *Global Change Biology*, 20 (9), 2886–2902.

Higgins, P. (2013). From sustainable development to carbon control: urban transformation in Hong Kong and London. *Journal of Cleaner Production*, 50, 56–67.

Höffken, J. I. and Limmer, A. (2019). Smart and eco-cities in India and China. *Local Environment*, 24 (7), 646–661.

HKSTP (2018a). Mission & core values. [Online] Available: https://www.hkstp.org/en/about-hkstp/the-corporation/mission-vision-core-values/ [Accessed 10 November 2020].

HKSTP (2018b). Veridian biotechnology limited. [Online] Available: https://www.hkstp.org/en/directory-list/Details/veridian-biotechnology-limited [Accessed 10 November 2020].

HKSTP (2018c). Pure innovation biotech limited. [Online] Available: https://www.hkstp.org/en/directory-list/Details/pure-innovation-biotech-limited [Accessed 10 November 2020].

Hong Kong Government (2012) The Gini coefficient of Hong Kong: Trends and interpretations. [Online] Available: http://www.hkeconomy.gov.hk/en/pdf/box-12q2-5-2.pdf [Accessed 10 November 2020].

Hong Kong Government (2016). Thematic household survey report no. 60. [Online] Available: http://www.statistics.gov.hk/pub/B11302602016XXXXB0100.pdf [Accessed 10 November 2020].

Hong Kong Government (2017). Housing in figures 2017. [Online] Available: http://www.thb.gov.hk/eng/psp/publications/housing/HIF2017.pdf [Accessed 10 November 2020].

Hong Kong Government (2018a). Hong Kong Smart City Blueprint. [Online] Available: https://www.smartcity.gov.hk/ [Accessed 10 November 2020].

Hong Kong Government (2018b). Gov't to boost R&D, AI. [Online] Available: https://www.news.gov.hk/eng/2018/09/20180928/20180928_154819_671.html [Accessed 10 November 2020].

Hong Kong Government (2018c). HK to focus on AI tech. [Online] Available: https://www.news.gov.hk/eng/2018/10/20181013/20181013_183341_019.html [Accessed 10 November 2020].

Hong Kong Government (2019). HK a biotech centre. [Online] Available: https://www.news.gov.hk/eng/2019/05/20190529/20190529_152626_328.html [Accessed 10 November 2020].

Hong Kong Planning Department (2007). Hong Kong 2030. [Online] Available: https://www.heritage.org/index/ranking [Accessed 10 November 2020].

IBM (2017). Creating a smarter hotel. [Online] Available: http://www-07.ibm.com/hk/smb/smarter_hotel/ [Accessed 10 November 2020].

IBM (2018). Getting started to smarter retailing. [Online] Available: http://www-07.ibm.com/hk/smartercommerce/retailing/customeracquisition.html [Accessed 10 November 2020].

Index of Economic Freedom (2020). Country rankings. [Online]. Available: https://www.heritage.org/index/ranking [Accessed 10 November 2020].

Jirón, P., Imilán, W. A., Lange, C. and Mansilla, P. (2020). Placebo urban interventions: Observing smart city narratives in Santiago de Chile. *Urban Studies*, 0042098020943426.

Karvonen, A., Cugurullo, F. and Caprotti, F. (Eds.) (2018a). *Inside smart cities: Place, politics and urban innovation.* Routledge, London.

Karvonen, A., Cugurullo, F. and Caprotti, F. (2018b). Introduction: Situating smart cities. In Karvonen, A., Cugurullo, F. and Caprotti, F. (Eds.). *Inside Smart Cities: place, politics and urban innovation.* Routledge, London.

Karvonen, A., Cugurullo, F. and Caprotti, F. (2018c). Conclusions: The long and unsettled future of smart cities. In Karvonen, A., Cugurullo, F. and Caprotti, F. (Eds.). *Inside Smart Cities: place, politics and urban innovation.* Routledge, London.

Kitchin, R. (2015). Making sense of smart cities: Addressing present shortcomings. *Cambridge Journal of Regions, Economy and Society*, 8 (1), 131–136.

Kitchin, R. (2016). *Getting smarter about smart cities: Improving data privacy and data security.* Data Protection Unit, Department of the Taoiseach, Dublin, Ireland. Available: http://mural.maynoothuniversity.ie/7242/1/Smart.

Lai, K. (2012). Differentiated markets: Shanghai, Beijing and Hong Kong in China's financial centre network. *Urban Studies*, 49 (6), 1275–1296.

Long, H., Liu, Y., Hou, X., Li, T. and Li, Y. (2014). Effects of land use transitions due to rapid urbanization on ecosystem services: Implications for urban planning in the new developing area of China. *Habitat International*, 44, 536–544.

LSE Cities (2011). Hong Kong's housing shame. [Online] Available: https://lsecities.net/media/objects/articles/hong-kong/en-gb/ [Accessed 10 November 2020].

Machado Junior, C., Ribeiro, D. M. N. M., da Silva Pereira, R. and Bazanini, R. (2018). Do Brazilian cities want to become smart or sustainable?. *Journal of cleaner production*, 199, 214–221.

March, H. and Ribera-Fumaz, R. (2018). *Barcelona: From corporate smart city to technological sovereignty.* In Karvonen, A., Cugurullo, F. and Caprotti, F. (Eds.). *Inside Smart Cities: place, politics and urban innovation.* Routledge, London.

Mathews, G. (2011). *Ghetto at the center of the world: Chungking Mansions, Hong Kong.* University of Chicago Press, Chicago.

McLean, A., Bulkeley, H. and Crang, M. (2016). Negotiating the urban smart grid: Socio-technical experimentation in the city of Austin. *Urban Studies*, 53 (15), 3246–3263.

Milano Smart City (2017). Smart Milano. [Online] Available: http://www.milanosmartcity.org/joomla/ [Accessed 10 November 2020].

Mouton, M. (2020). Worlding infrastructure in the global South: Philippine experiments and the art of being 'smart.' *Urban Studies*, 0042098019891011.

Nkulu, C. B. L., Casas, L., Haufroid, V., De Putter, T., Saenen, N. D., Kayembe-Kitenge, T., Obadia, P.M., Wa Mukoma, D.K., Lunda Ilunga, J.M., Nawrot, T.S., Luboya Numbi, O., Smolders, E. and Nemery, B. (2018). Sustainability of artisanal mining of cobalt in DR Congo. *Nature sustainability*, 1 (9), 495–504.

Offenhuber, D. (2019). The platform and the bricoleur – Improvisation and smart city initiatives in Indonesia. *Environment and Planning B: Urban Analytics and City Science*, 46 (8), 1565–1580.

Odendaal, N. (2016). Getting smart about smart cities in Cape Town: Beyond the rhetoric. In Marvin, S., Luque-Ayala, A. and McFarlane, C. (Eds.). *Smart urbanism: utopian vision or false dawn?* Routledge, London.

Parks, D. (2019). Energy efficiency left behind? Policy assemblages in Sweden's most climate-smart city. *European Planning Studies*, 27 (2), 318–335.

Peng, X., Shi, D., Guo, H., Jiang, D., Wang, S., Li, Y. and Ding, W. (2015). Effect of urbanisation on the water retention function in the Three Gorges Reservoir Area, China. *Catena*, 133, 241–249.

Qiu, B., Li, H., Zhou, M. and Zhang, L. (2015). Vulnerability of ecosystem services provisioning to urbanization: A case of China. *Ecological Indicators*, 57, 505–513.

Raco, M. and Gilliam, K. (2012). Geographies of abstraction, urban entrepreneurialism, and the production of new cultural spaces: The West Kowloon Cultural District, Hong Kong. *Environment and Planning A*, 44 (6), 1425–1442.

Raco, M. and Street, E. (2012). Resilience planning, economic change and the politics of post-recession development in London and Hong Kong. *Urban Studies*, 49 (5), 1065–1087.

Ranchod, R. (2020). The data-technology nexus in South African secondary cities: The challenges to smart governance. *Urban Studies*, 0042098019896974.

Reuters (2019). Exclusive: The Chief Executive 'has to serve two masters' – HK leader Carrie Lam – full transcript. [Online] Available: https://www.reuters.com/article/us-hongkong-protests-lam-transcript-excl/exclusive-the-chief-executive-has-to-serve-two-masters-says-hong-kong-leader-carrie-lam-full-transcript-idUSKCN1VX0P7 [Accessed 10 November 2020].

SCMP (2018). Universities warn that Hong Kong must 'innovate or die' amid growing economic challenges from Greater Bay Area cities. [Online] Available: https://www.scmp.com/news/hong-kong/hong-kong-economy/article/2146076/universities-warn-hong-kong-must-innovate-or-die [Accessed 10 November 2020].

Shen, J. (2008). Hong Kong under Chinese sovereignty: Economic relations with mainland China, 1978–2007. *Eurasian Geography and Economics*, 49 (3), 326–340.

Shelton, B., Karakiewicz, J. and Kvan, T. (2013). *The making of Hong Kong: from vertical to volumetric*. Routledge, London.

Shelton, T., Zook, M. and Wiig, A. (2015). The 'actually existing smart city.' *Cambridge Journal of Regions, Economy and Society*, 8 (1), 13–25.

Smart City Wien (2019) Framework strategy. [Online] Available: https://smartcity.wien.gv.at/site/en/ [Accessed 10 November 2020].

Spicer, Z., Goodman, N. and Olmstead, N. (2019). The frontier of digital opportunity: Smart city implementation in small, rural and remote communities in Canada. *Urban Studies*, 0042098019863666.

Tam, C. M., Deng, Z. M., Zeng, S. X. and Ho, C. S. (2000). Quest for continuous quality improvement for public housing construction in Hong Kong. *Construction Management & Economics*, 18 (4), 437–446.

The Economist Intelligence Unit (2018). The Economist Intelligence Unit's Democracy Index. [Online] https://www.economist.com/graphic-detail/2019/01/08/the-retreat-of-global-democracy-stopped-in-2018 [Accessed 10 November 2020].

The Economist Intelligence Unit (2019). Business Environment – Hong Kong. [Online] http://country.eiu.com/article.aspx?articleid=398487423&Country=Hong%20Kong&topic=Business&subtopic=Business+environment&subsubtopic=Rankings+overview [Accessed 10 November 2020].

The Guardian (2013). Hong Kong's cubicle apartments: could you live like this? [Online] Available: http://www.theguardian.com/world/gallery/2013/feb/22/hong-kong-flats-tiny-cubicles [Accessed 10 November 2020].

The Guardian (2017). My week in Lucky House: the horror of Hong Kong's coffin homes. [Online] Available: https://www.theguardian.com/world/2017/aug/29/hong-kong-coffin-homes-horror-my-week [Accessed 10 November 2020].

Tironi, M. and Valderrama, M. (2018). Unpacking a citizen self-tracking device: Smartness and idiocy in the accumulation of cycling mobility data. *Environment and Planning D: Society and Space*, 36 (2), 294–312.

Trencher, G. and Karvonen, A. (2019). Stretching 'smart': Advancing health and well-being through the smart city agenda. *Local Environment*, 24 (7), 610–627.

van den Brink, S., Kleijn, R., Sprecher, B. and Tukker, A. (2020). Identifying supply risks by mapping the cobalt supply chain. *Resources, Conservation and Recycling*, 156, 104743.

Wan, L., Ye, X., Lee, J., Lu, X., Zheng, L. and Wu, K. (2015). Effects of urbanization on ecosystem service values in a mineral resource-based city. *Habitat International*, 46, 54–63. doi:690.

Wiig, A. and Wyly, E. (2016). Introduction: Thinking through the politics of the smart city. *Urban Geography*, 37, 485–493.

Wiig, A. (2018). Secure the city, revitalize the zone: Smart urbanization in Camden, New Jersey. *Environment and Planning C: Politics and Space*, 36 (3), 403–422.

Wong, T. K. Y. and Wan, P. S. (2009). Lingering environmental pessimism and the role of government in Hong Kong. *Public Administration and Development*, 29 (5), 441–451.

World Bank (2015). CO2 emissions – Hong Kong SAR. [Online]. Available: https://data.worldbank.org/indicator/EN.ATM.CO2E.PC?locations=HK [Accessed 10 November 2020].

World Wide Fund (WWF) (2015). Marine – Hong Kong. [Online]. Available: http://www.wwf.org.hk/en/whatwedo/conservation/marine/ [Accessed 10 November 2020].

PART III

The apocalypse

'It was on a dreary night of November that I beheld the accomplishment of my toils. My candle was nearly burnt out when, by the glimmer of the half-extinguished light, I saw the dull yellow eye of the creature open.'

(*Frankenstein, Chapter 5*)

PART III

The apocalypse

6

URBAN EQUATIONS COME ALIVE

Introduction

As an experiment begins, it must also come to an end. Preliminary studies are conducted, the laboratory is set up, the methodology is tested and the experiment is finally carried out. Then, it could be a matter of minutes, hours or years, but eventually the act of experimenting stops. This does not mean of course that the last stage of an experimental project is a sudden conclusion which truncates the experimentation, freezing its actors and closing its context. Clean-cut ends are rarely the case. Experiments can have long-standing consequences and resonance, and what the experimenters have triggered might continue to exist after a given project is formally terminated. One might say that this is actually the moment when the real experiment starts. The subject of the experiment is not limited anymore by the confined boundaries of the methodology and the research facilities. It is free to interact with the outside world.

Mary Shelley's *Frankenstein* offers an example of these dynamics. After having studied the theory, prepared the methodology and organized the laboratory, Victor Frankenstein conducts the experiment. His endeavour lasts for two years. This is a precise time during which the young scientist experiments with the development of a human being, trying to capture not simply the secret of life, but the essence of the ideal form of life. The aim is to create the perfect human being. After this period, Victor, under the impression that his goal has been achieved, terminates the experiment (intended as the act of experimenting), but he does not end what he has created. The experiment, as the product of experimentation, is alive. Its eyes are open. It moves. It interacts. It has agency and its actions will have impacts and repercussions.

When it comes to experimental urbanism, the situation is more complex. The city itself is an open laboratory which does not have close and precisely defined boundaries. There is a plethora of actors involved. Some of them, like architects and planners, might quickly leave the site of experimentation after their job is done. Others, such as politicians and policy-makers, could spend the rest of their life in the same city where the experimentation took place. Above all, as remarked in different strands of geography and urban studies, cities are not static artefacts, but rather processes (see, for instance, Batty, 2018; Kaika, 2005; Keil, 2005). As such, they do not have an expiration date and a clear end. Cities keep changing and hence living, as part of long-term social, political, economic, and environmental transformations. By the same token, experimental cities cannot be simply terminated. When an urban experiment is metabolized by the city hosting it, its components join an indefinite process of urban development which can be sustained for centuries.

Nonetheless, despite the peculiarity of the development of cities, there is a key aspect which experimental urbanism shares with the example of *Frankenstein*. At some point, the act of experimenting ends. Whether it is carried out by a single lone genius or a team of urban planners, engineers and policy-makers, experimentation, at a certain time and in a certain place, eventually stops. At that time and at that place the product of experimentation can be found: a product which its creators have to face together with the consequences of what they have done. In the case of *Frankenstein*, it is sheer horror. The horror that Victor feels when he finally sees the result of his experiment (the horrendous yellow-eyed creature in front of him), and the horror of the crimes that the monster perpetuates. In the case of experimental urban projects, such as eco and smart-city initiatives, the outcomes and repercussions of their development will be the focus of this chapter.

The following sections build upon the empirics discussed in the second part of the book, to *unveil* the results of experimental eco and smart-city projects. The choice of the term 'unveiling' serves here a twofold purpose, and resonates with the concept of the *apocalypse* which is the leitmotif of the third and final part of the book. As noted by Gray (2008: 5), when the word 'apocalypse' is employed in common speech, it usually evokes images and feelings of a catastrophic event, but this is not its original meaning: 'in biblical terms it derives from the Greek word for unveiling.' Etymologically, this word is ἀποκάλυψις (apokálypsis), from the union of ἀπό and καλύπτω, and it signifies the act of removing a veil. However, the two interpretations do not necessarily exclude each other. In Mary Shelley's novel, for example, Victor removes the veil that covers his creation, thereby unveiling the results of the experiment. Subsequently, once the monster is free to interact with the outside world, the consequences are indeed catastrophic. Frankenstein's creature harms most of the people whom it encounters, killing several of them, and putting the whole of humanity (so his creator believes) in danger. Victor's experiment thus triggers what in common speech would be referred to as an apocalypse. For reasons that Victor struggles to understand, the experiment is not sustainable. It does not look like anything that the Doctor had in mind and, above all, its behaviour puts in danger what is dearest to Victor. This chapter unveils the outcomes of the urban experiments examined earlier, questioning their sustainability and interrogating the reasons why, like Frankenstein's monster, they fail to be sustainable.

Four sections form the rest of the chapter. Each one of them picks a fundamental problem behind Victor's tragic experiment, and then discusses it in relation to the case studies analyzed in Chapters 4 and 5. First, the narrative delves into the fragmentation characterizing the shape and functions of Abu Dhabi's eco-city project and Hong Kong's smart interventions. Here the chapter underlines how the two urban experiments are composed of elements which are individually perfect, but often incompatible with each other, stressing that this incompatibility is a source of unsustainability. Second, the discussion moves to the behaviour of the experiments. The focus shifts to what Frankenstein's monster and, likewise, the Masdar City project and Hong Kong's smart-city initiatives actually do once they become operative. The chapter claims that their actions and repercussions are catastrophic, because several gaps in knowledge undermine the way the experiments understand what should be sustained and how. To explore this critical issue in-depth, the book develops and proposes the concept of *urban equations*, in the attempt to visualize the sustainability themes that have been processed, misinterpreted or, worse, ignored in the formulation of the experimental eco and smart-city projects. The third and final problem that is observed is the severe deregulation of both the Masdarian experiment and Hong Kong's smart urbanism which, like Frankenstein's creature, are not controlled by their makers. The experiments are left alone, unsupervised, with nobody checking their sustainability and being responsible for their performance, particularly from an environmental and social point of view. The chapter ends by suggesting that a positive finale is still possible. Urban experiments are not born monsters: they can become monstrous. Their flaws are predictable, and the last section suggests alternative practises and understandings of experimental urbanism, oriented toward sustainability.

Frankenstein urbanism

After two years of strenuous work, Victor Frankenstein decides to finish his experiment. He has done everything that was part of his plan, and it is time for him to examine the outcomes of the project. He is alone in his laboratory and, with the light of a candle, he reveals the features of the thing that he has created. This is what the Doctor sees, as it is painfully described through his very words:

> How can I describe my emotions at this catastrophe, or how delineate the wretch whom with such infinite pains and care I had endeavored to form? His limbs were in proportion, and I had selected his features as beautiful. Beautiful! Great God! His yellow skin scarcely covered the work of muscles and arteries beneath; his hair was of a lustrous black, and flowing; his teeth of a pearly whiteness; but these luxuriances only formed a more horrid contrast with his watery eyes, that seemed almost of the same colour as the dun-white sockets in which they were set, his shrivelled complexion and straight black lips.
>
> *(Shelley, 2013: 45)*

Victor is not only scared of the creature next to him, he is also deeply disgusted by what his eyes are seeing: a monster. The horror filling him is such that he cannot even look at the product of the experiment. Unable to tolerate the aspect of his creation, Victor rushes out of the room. It took him two years of experimentation, during which the Doctor sacrificed much of his life. Two years of solitude, wholly dedicated to the experiment and, yet, it takes only a couple of seconds for his dream to turn into a nightmare. He quickly realizes that something has gone wrong, but what exactly did not work is not obvious.

The reaction of Victor is not surprising per se. Every person who encounters Frankenstein's creature is horrified by it because of its appearance. In contemporary culture, the being created by Frankenstein is considered to be an archetypical monster and a prime instance of horror, largely due to how it looks. In Mary Shelley's novel, every time someone sees it or when people in today's society think about it, the image that is conjured up is that of an ugly, disgusting and horrid creature. What is surprising, however, is that this image does not exactly correspond to the description initially made by Victor when he first looked at the creature. An accurate reading of the Doctor's testimony reveals an unexpected truth. The Doctor clearly says that the features of the man that he has made are beautiful. He points out that his limbs are in proportion, which is a characteristic of the human body normally associated with beauty. Victor goes on, noticing how his hair is of a lustrous black and his teeth of a pearly whiteness. Looking at each of the elements of the body in front of him, the Doctor unmistakably sees beauty and luxury, rather than horror. This is because his original plan was to create a creature that was immortal and also beautiful. To this end, as part of the experiment, he had carefully selected different pieces, making sure that each one was perfect and beautiful. He had found and then fitted together perfect limbs, muscles, hair and teeth. Yet, the final product is dreadful and, despite the beauty of the single components, the appearance of the man facing Victor is that of a monster.

In order to understand what actually makes Frankenstein's creature look like a monster, the initial description given by Victor is again illuminating. As the Doctor notes, the problem lies in the *contrast*. There is nothing wrong with the single pieces that form the end product of the experiment. Individually they are perfect. It is the contrast or, in other words, the lack of connection among the heterogeneous elements that Victor has forced together, which causes the experiment to fail. On these terms, the problem relates to the shape of the experiment. It is made of incongruous components which, when next to each other, create a horrendous whole. Seen from this perspective, the issue is almost purely aesthetical in nature, and one might wonder why Victor did not notice immediately such a flaw while he was conducting the experiment. In reality, he did. Reflecting upon the days before the end of the project, Victor admits:

'I had gazed on him while unfinished; he was ugly then, but when those muscles and joints were rendered capable of motion, it became a thing such as even Dante could not have conceived' (Shelley, 2013: 46).

There is thus a deeper problem undermining the experiment, which is not just about its shape. The elements forming the creature are incompatible. Not only do they look horrid when combined but, more fundamentally, they do not function well together. The monster becomes such the moment it is alive. It is horrendous when its parts start to function. Then Victor truly realizes that those individually perfect pieces are not meant to be together because they do not work in harmony. The discord and tension that their union is generating make Frankenstein's creature worthy of hell.

The same problems can be observed in the two experimental urban projects examined earlier in the book. Masdar City is an urban experiment which fits together heterogeneous spaces, infrastructures and technologies in the attempt to create a single brand-new urban settlement. Although promoted by its developers as a master-planned project, in reality the construction of the city has not been disciplined by a coherent and precise master plan. This eco-city project has been cobbled together, like Frankenstein's monster, from disparate materials. In the Masdar City project, the developers have several business partners. Each one of them possesses the power to influence the shape of the city, even if this means going directly against the vision of the urban planners who are officially in charge of the layout of the built environment. Every business partner adds a new building or a new urban technology to Masdar City without adhering to an overarching master plan. The result is, in practise, a process of urban development in which different actors add different pieces of the city in an uncoordinated and individualistic manner. The outcome of this fragmented process is, like in *Frankenstein*, a creature whose elements are not fully compatible with each other.

An example of this patchy urbanism is the implementation of two iconic components of Masdar City: the Personal Rapid Transit (PRT) advanced by Foster+Partners and Mitsubishi's electric cars. These two elements, at the core of the transport portfolio of the new city, are radically different in relation to their philosophy and infrastructure. The PRT, as a system of shared driverless cars, is meant to support public transport. The electric car sold by Mitsubishi, instead, is in line with the culture of private ownership. Moreover, the PRT functions mostly on a separate level of Masdar City (the undercroft) in order to make the whole surface of the city completely pedestrian and car-free. Mitsubishi's technology is against that and can operate only in a city where cars are allowed and have access to most urban spaces. Such radical differences make these two pieces of Masdar City incompatible in terms of urban design. Cars cannot fulfil their function in the narrow pedestrian streets originally envisioned by Foster+Partners and, of course, they cannot exist in a car-free city: they need broad vehicle lanes and a network of roads, crisscrossing the entire settlement. Figure 6.1 and Figure 6.2 show the Masdarian urban spaces designed according to the *car-free* ideal. Figure 6.3 shows how the built environment of Masdar City has been designed to accommodate cars. The two urban designs are substantially different from each other and, yet, they are geographically and chronologically very close. They belong to areas that are adjacent and were developed during the same phase of the Masdar City project.

FIGURE 6.1 The city centre of Masdar City. The first area that was built and opened to the public. It is fully pedestrian and reflects Foster+Partners' original vision of a car-free city. Before the master plan was disrupted by the Masdar Initiative's partnership with Mitsubishi, this is how the whole city was supposed to look like.

Source: Gianfranco Serra Photography

It is important to note that, individually, both the PRT system and the fleet of Mitsubishi's electric cars function well, performing up to the expectations of their creators. The problem lies in the contrast between them, inasmuch as the diffusion of the latter hinders the development of the former. In fact, when in 2010 the developers of Masdar City decided to integrate electric vehicles into the transport system of the new settlement, they had to reject the *car-free* concept and stop the implementation of the PRT together with the city's undercroft. From a sustainability perspective, the consequences of this inner tension have been significant. With the PRT limited to 10 per cent of the settlement, Masdar City has become a car-centric city. Streets are not being built anymore as narrow lanes which maximize natural shading and air circulation, but as broad vehicle lanes which facilitate the transit of cars and trucks. This urban design increases the perceived temperature in the built environment which, in turn, intensifies the dependency on air-conditioning and cars. In a place like Masdar City incapable to fully power itself via renewable energy, such dependency is sustained by the burning of oil and gases. Moreover, in addition to this severe environmental problem, the social benefits provided by a pedestrianized settlement, in terms of health, wellbeing and social interactions, have been eliminated

FIGURE 6.2 A typical street in the centre of Masdar City, designed according to
Foster+Partners' original vision. Every element of the built environment is
constructed and shaped to facilitate the passage of pedestrians. It is not a
space for cars.

Source: Antonio Mannu

by the spreading of cars. Looking at the shape of the new Emirati city, it is clear that
the PRT and Mitsubishi's cars were never meant to be together, but their incompatibility became evident only when they began to function next to each other.

In Hong Kong, the regional smart-city agenda is presented by the government
as a blueprint, but its components do not actually follow any cohesive plan of
action. In the neoliberal context where smart urbanism is being practised, private
developers take the lead, regenerating old buildings and creating new high-tech
ones, across a largely privatized landscape. They build in an individualistic manner,
and pursue their agenda without thinking of Hong Kong as a whole, instead
focusing on the single parcel of land that they control. The emphasis is thus on the
single pieces, not on their sum. No one is coordinating the many smart interventions that take place in the region and, as a result, smart buildings rise in isolation
from the rest of the city, bereft of a connection with the surrounding built environment. Individually, they function well. Because of the many smart technologies
permeating them, they can save energy, reduce waste, decrease carbon emissions,
increase safety and boost the business activities of their owners. However, they are
inaccessible to most of the population and their very presence exacerbates the
already deep divide between the rich and the poor. Only high-income workers
who have access to adequate housing can afford them. Whoever cannot pay for
smart is left outside, and has to opt for what is often a low-quality accommodation

FIGURE 6.3 The evolution of Masdar City after the inclusion of cars. The narrow and shaded pedestrian streets have disappeared, except in the city centre which represents the past of the Emirati eco-city project. Broad vehicle lanes hit by the scorching sun are the present and future of Masdar City.

Source: Gianfranco Serra Photography

in a run-down estate. Those excluded from smart urbanism, the outsiders, do not have to go very far since the alternative is not located in another city. Within Hong Kong, smart and non-smart urban spaces frequently share the same district and, at times, even the same road (see Figures 6.4 and 6.5). Yet, they are not connected in a homogeneous manner. These two spatial typologies belong to different social classes and are characterized by a distinct infrastructure: a dissonant urban condition that the local planning council has no intention of harmonizing.

The only two districts where the government is consistently and thoroughly rolling out smart interventions also present a considerable degree of fragmentation. As Figure 5.1 has shown, the Hong Kong Central Business District Two and the Hong Kong Science and Technology Park are patches of *smart* affixed to much larger pieces of urban fabric. The Science Park, for instance, is geographically detached from the rest of Hong Kong. In practise, it consists of an isolated built environment where private companies test experimental smart technologies. Its location is remote, and its function serves mostly the purpose of an elite of business enterprises which include the Chinese tech giant Alibaba. Inside the Park, technological innovation is pursued in line with the political economy of Hong Kong, but without a holistic vision of the city. The focus is largely on technology crafted to improve single homes or, at most, single buildings. The technological production of the Science Park, therefore, does not foster the development of the city as a

FIGURE 6.4 The genesis of One Taikoo Place. This is a smart building constructed in 2018 and located in Quarry Bay (Hong Kong Island). It is Hong Kong's first building to be fully operated by an artificial intelligence: an innovative feature for which One Taikoo Place is hailed by the developers as the pinnacle of smart urbanism in Hong Kong. The project was financed and developed by a private company called Swire Properties, while the artificial intelligence (named Neuron) was provided by another private company, ARUP.

Source: Wikimedia Commons

whole, but only of some of its individual pieces. Moreover, it does not support the preservation of the natural environment. Its buildings and infrastructures are disconnected from local ecosystems. The construction of the Park meant the eradication of a portion of the flora and fauna of Hong Kong because, by design, the structures of steel, glass, plastic and concrete forming this district could not coexist

FIGURE 6.5 On King's Road, the same road where One Taikoo Place stands, is also located what locals have nicknamed 'Monster Building,' an old and hyper-dense housing estate. In Hong Kong, the smartest building and the most infamous building are just 300 metres apart from each other.

Source: Wikimedia Commons

in the same space with the pre-existing biomass. The smart built environment was not compatible with the natural environment, and the development of the former caused the loss of the latter.

Flawed urban equations

The second reason why Frankenstein's creature is commonly considered a monster is not due to its shape and form. Victor's experiment kills. It is not a simple matter of look then. Those muscles animated by electricity are strong and can break many things, including the life of people. The creature is far from being harmless. It is powerful and eager to interact with the rest of the world. In the story narrated by Mary Shelley, it soon leaves the laboratory and starts to discover the surrounding environment, exploring cities, travelling off the beaten track and engaging with several characters. Many of them die by the hands of the creature and the monster is quickly seen as such: a reckless murderer and enemy of humanity. The Doctor has no doubt regarding the antagonistic nature of his creation and declares:

My abhorrence of this fiend cannot be conceived. When I thought of him I gnashed my teeth, my eyes became inflamed, and I ardently wished to extinguish that life which I had so thoughtlessly bestowed. When I reflected on his crimes and malice, my hatred and revenge burst all bounds of moderation.

(Shelley, 2013: 87)

When he is proclaiming his hatred for the monster, the Doctor is not thinking of its appearance, but rather of its actions. He does not have in mind an ill-shaped man. His thoughts are focused on the crimes that the creature has committed: on the people whom Victor cared about and the monster brutally killed. The monstrosity and, above all, the failure of Victor's experimental project is thus deeply connected to the behaviour of the experiment and, in essence, to what the creature actually does. In *Frankenstein*, the creature performs a number of actions, but given that killing human beings, in particular, is the activity marking it as a monster, it is important to understand the reason behind this specific type of behaviour.

Frankenstein's monster is an exceptionally intricate character and its behaviour, at times ferocious, at times gentle, cannot be explained on the basis of a single factor. However, it is possible to extrapolate from its actions an underpinning problem: there is a flaw in how the creature perceives reality and this flaw compromises its behaviour. The first murder, for instance, is not intentional. The monster is initially unaware of its immense strength and does not know that human beings are fragile creatures whom its big hands can so easily harm. It is not even fully familiar with the concepts of life and death in the first place, and its actions are catastrophic because it cannot foresee their repercussions. Basic cause–effect relationships escape its understanding which is full of gaps in knowledge and suffers from limited information about the world and the complex rules that govern its physics and societies. Even later in Shelley's story, when the creature perfectly understands that it is capable of killing, it does not hesitate to harm Victor's friends, because it believes that this is the right thing to do. Such belief, alas, does not take into account the disastrous escalation that the murders will trigger, leading eventually to a complete tragedy. Its catastrophic behaviour is, therefore, repeatedly driven by a problematic idea of what is sustainable and what is unsustainable.

Reflecting upon the case studies previously examined, similar problems can be found in how experimental urban projects understand and then approach sustainability. The way the Masdar City project and Hong Kong's smart-city agenda formulate and address urban sustainability is flawed, due to many severe gaps in knowledge and misunderstandings regarding what ideas and actions are supposed to be sustainable. First, in the case of Masdar City, it is clear that, unlike Frankenstein's creature, the new Emirati settlement does not have a brain or a mind. Hence, it does not think. This, however, does not mean that the project does not acquire and process knowledge, ultimately developing an understanding of the meaning of urban sustainability. Such a process is not carried out directly by the city itself. It is executed indirectly by the urban developers who decide what ideas will be at the core of the project, will subsequently drive its implementation and, ultimately, will influence the actual performance of the experiment. From a conceptual point of view, the understanding of urban sustainability which the developers of Masdar City have formulated for the new city, can be visualized as a simple equation.

$$A = B$$

The two sides are equal to each other. More specifically, the focus here is on how urban sustainability is formulated or, in other words, on what urban sustainability is equal to. The concept of urban sustainability is represented on the left side of the equation, with B symbolizing instead the diverse factors which together can potentially lead to a sustainable urban development.

$$Urban\ sustainability = B$$

Chapter 4 demonstrated that the Masdar Initiative formulated the eco-city project's understanding of urban sustainability in a very narrow way by focusing almost exclusively on business aspects, with the concept of profitability being equated to the concept of sustainability. For the developers, being profitable and being sustainable are synonyms. On these terms, the Masdar City project understands *sustainability* as *profitability*, and attempts to achieve this condition through the development of clean technologies. The technological portfolio of Masdar City focuses extensively on the reduction of carbon emissions, which is the main environmental theme of the Emirati eco-city project. When it comes to environmental sustainability, the Masdarian interpretation sees CO_2 as the enemy whose elimination, to paraphrase Swyngedouw (2010), will also eliminate climate change. Such conceptualization and stress on the importance of reducing CO_2 emissions is, of course, not a coincidence and relates to the fact that Masdar City's stakeholders can generate substantial profit by selling technology designed to fulfil the very idea that CO_2 must be eliminated. There are other environmental themes that feature in the Masdarian understanding of sustainability, such as water-waste reduction and (since 2019) sustainable agriculture for arid climates, but their prominence is considerably lower compared to the vast anti-CO_2 technological arsenal examined in Chapter 4. Similarly, the Masdar City project integrates social concerns and basic questions of justice, via the so-called *twenty per cent policy* meant to support the housing needs of the low-income share of Abu Dhabi's population. These concerns, however, occupy only a marginal role in the Masdarian formula for urban sustainability, as testified by the fact that it is a meagre 20 per cent of the city to be socially just, versus 80 per cent of elitist space. Besides, this policy has actually never been implemented.

The Masdarian understanding of what should be sustained and how, can be visualized as an equation which is not supposed to be taken in a literal way. An urban equation, such as the one expressed below, serves the purpose of visualizing how a certain experimental urban project like Masdar City, understands the concept of urban sustainability. In this sense, urban equations are a heuristic tool to quickly identify where the focus of developers and stakeholders is, what aspects of sustainable urban development have been taken into account and what themes have been dismissed.

$$A = \left[\left(\frac{Technological\ innovation}{Reducing\ carbon\ emissions}\right)Profitability\right]^{10}$$
$$+\ Water\ waste\ reduction + Sustainable\ agriculture$$
$$+\ Social\ housing$$

The case of Masdar City reveals that, in the Emirati eco-city project, urban sustainability (A) is understood as the sum of different elements, with an extreme focus on those that can generate profit and are related to the development of new commodifiable technologies. What is unprofitable is either undercounted or, worse, completely unaccounted. Social housing, for example, is an element of the Masdarian equation, but it has much less weight in comparison with the developers' intention to reduce the carbon emissions produced by the new city. In addition, it is evident that other important environmental and social themes, such as ecosystem services and gender equality, for instance, are entirely missing in the formula.

Second, in the case of the smart-city agenda of Hong Kong, similar problems emerge. As Chapter 5 explained, in Hong Kong smart urbanism is formulated in politico-economic terms, with smart-city initiatives feeding into a regional strategy of economic growth. The elements that appear in the formula are exclusively those that are connected to the economic sectors in which the government wants to invest. Following this line of thought, for Hong Kong's policy-makers, being *smart* is about Information and Communication Technology (ICT), biotechnology and, more recently, artificial intelligence (AI). These are the three sectors of the local economy that can attract investment the most, thus creating new jobs and yielding stable economic activities: such is the belief of the makers of *Hong Kong Smart City Blueprint*, which then permeates into how regional smart interventions approach *smart*. Because of this politico-economic understanding focused on the cultivation of novel high-tech strands of the economy, it is not a surprise that preserving the biodiversity of the region, for example, is not considered to be *smart*. Hence the development of smart districts which are capable of attracting multinationals, generating new business and facilitating the genesis of novel biotechnologies and artificial intelligences, while being completely incapable of sustaining the natural environment. At a smaller scale, the same mentality leads to the creation of automated smart buildings which have the intelligence to comprehend their own energy performance, but cannot speak the language of ecology and, therefore, fail to connect with the surrounding bioregion. The underlying problem is in the way *smart* has been formulated. Hong Kong's smart urban spaces are insensitive to ecological issues, because ideas of urban ecology were not taken into account by the local policy-makers when they created a smart-city agenda. The result of this lacuna is a simplistic urban equation.

$$Smart = ICT + Biotechnology + AI$$

Oversimplification, like the one manifested by the smart-city agenda of Hong Kong is dangerous from a sustainability point of view. There is widespread agreement in the scientific community on the nature of urban sustainability as a complex condition whose achievement necessitates a delicate balance among many interrelated social, environmental and economic factors (James, 2014; Elmqvist et al., 2019; Long and Rice, 2019; Robertson, 2017; Verma and Raghubanshi, 2018; Yan et al., 2018). In this sense, oversimplifying the spectrum of themes taken into account in the development of an experimental urban project is a problem. The formulation of urban equations that are based upon just a handful of elements heavily skewed toward economic matters, will not ultimately produce sustainable cities, inasmuch as crucial socio-environmental issues will be ignored. Going back to the *Frankenstein* analogy, the two case studies fail to be sustainable because, like the monster, they do not fully understand and comprehend what is around them. Frankenstein's creature has not been told that human beings can die and that its strength can kill them. Likewise, Hong Kong's smart districts have not been given any ecological awareness by their makers, and so their very presence damage the local ecology. Similarly, Masdar City concentrates its efforts on the reduction of carbon emissions, ignoring other important environmental problems because there is a gap in the city's understanding of environmental sustainability. Moreover, when the monster starts to intentionally kill Victor's friends, it is because, according to its ideas, this agenda is correct. In figurative terms, both Masdar City and Hong Kong suffer from problematic ideas about what the right thing to do is. The Masdar City project, for instance, pursues an agenda which favours an elite (the royal family of Abu Dhabi and the partners of the Masdar Initiative), in line with the directions of the developers. Hong Kong's smart-city initiatives, for example, strive to create premium office spaces for multinationals (thus ignoring the dire housing crisis which affects the region), since in the mind of their creators this is the right plan of action.

The flawed urban equations which underpin the genesis and implementation of the Masdarian eco-city project and Hong Kong's smart-city agenda, do not capture the complexity of urban sustainability. This limitation puts the emphasis on critical gaps in the formulation of experimental urban projects, as well as on how ideas of social and environmental sustainability have been (mis)interpreted to fit an economic logic. Chapter 4 exposed the Masdar Initiative's interpretation of social sustainability as *customer satisfaction*, while Chapter 5 has elucidated how in Hong Kong *sustainable* has to rhyme with *neoliberal*. Ideas, however, are not simply abstract categories of thought. Their very existence depends upon who conceives and embraces them. Questioning an idea, therefore, is a step which naturally leads to questioning the bearer of that idea, but this is an inquiry which should be approached with care. Frankenstein's monster, for instance, was abandoned by Victor right at the end of the experiment. Whether the creature should be responsible for its sins or the responsibility lies on its maker is an open question to which the chapter will now turn.

Alone in the dark

The third reason why Victor's experiment fails is that the product of experimentation is left alone, uncontrolled and without any sort of supervision. This is also the reason why the Doctor can be seen as the real monster. Victor abandons his creature immediately, even though he is *de facto* his father. He takes no responsibility and avoids giving an education to the man that he has poorly created when the miserable new born needs him the most. Afterwards, Victor fails to empathize with the creature, while the latter is desperately seeking his help. He constantly pursues his own agenda and only really starts to worry about the experiment when the monster is killing members of his personal network. It is then that he finally decides to intervene and deal with the repercussions of his experimental project by trying to destroy it. Because of Victor's negligence, it is thus plausible to assume that the creature was not born a monster but became one due to its creator. The same theory is defended by the creature itself when it confronts its maker:

'Believe me, Frankenstein, I was benevolent. My soul glowed with love and humanity, but am I not alone, miserably alone? You, my creator, abhor me. What hope can I gather from your fellow creatures, who owe me nothing?' (Shelley, 2013: 96).

Frankenstein would arguably not have been a horror story and certainly not a tragedy if Victor had been more present and responsible. The Doctor, for example, could have checked the development of his creature, monitoring its behaviour as well as the ideas that it was cultivating. He could also have prepared a backup plan to stop the monster in case it started to harm innocent people. In reality, instead, Victor is never there when the creature begins to painfully learn about the meaning and implications of being alive, and later when the monster starts to kill. Frankenstein's creature is always alone in the dark. The darkness of the laboratory where it is abandoned and the darkness of ignorance that obscures its understanding of an unknown and hostile real world. Above all, the creature is alone in the sense that it is uncontrolled. Nobody is directly controlling its actions or being responsible for them. It is precisely this lack of control and responsibility that is the cause of the escalation of the tragic events unfolded in Shelley's story.

The fate of the Masdar City project and Hong Kong's smart-city agenda resembles the fate of Frankenstein's creature. Like the monster, both urban experiments are alone because their performance is largely unsupervised. Many of the repercussions of the two experimental urban projects are unchecked, since their developers, like Victor Frankenstein, pay little or no attention to what their creations actually do once operative. The creature of the Masdar Initiative, for example, is constantly monitored from an economic perspective. Chapter 4 pointed out that the development of the Masdar City project depends upon detailed economic analyses and business strategies which indicate what is profitable, with the developers adjusting the shape and technological portfolio of the new city accordingly. Conversely, in relation to broader aspects of sustainability concerning the environmental and social dimensions of the project, the growth of Masdar City is uncontrolled. There is a building regulatory system supposed to discipline the construction of the Masdarian built environment and check its sustainability. However, in reality, this is a façade carefully crafted by the Masdar Initiative to carry out undisturbed its many business projects. More specifically, the regulatory system is called Estidama which is the Arabic word for *sustainability* (Cugurullo, 2013). Estidama is mandatory in Abu Dhabi and is described by the local Urban Planning Council (UPC) as a 'framework for measuring sustainability' (UCP, 2015 no page). The core element of Estidama is the so-called Pearl Rating System which provides several parameters, such as carbon emissions, water consumption and energy waste, through which the UPC is meant to assess the sustainability of Abu Dhabi's new buildings and cities. The mechanism is straightforward. In order to pass the Estidama test, an urban project (regardless of its scale) needs one Pearl, unless it is led by the government, in which case two Pearls are needed. The UPC can grant up to five Pearls, thereby assessing the extent to which a given urban space is sustainable.

While, in theory, Estidama could be a useful tool to control the sustainability of the Emirati built environment, the practise of this regulatory system shows otherwise. Estidama and the Masdar City project were developed around the same time, when the government of Abu Dhabi decided to revamp its urban agenda. More precisely, according to a manager from the UPC, 'Masdar City, as an idea, came before Estidama. Estidama started in summer 2007. Masdar City was on the drawing board at least a year before then.' The problematic aspect of the history of Estidama is that the actors behind its inception are those whom the system was eventually going to evaluate: it was the Masdar Initiative itself that designed the initial mechanics of the Pearl Rating System, while planning the mechanics of Masdar City. 'There is a lot of crossover and affinity between the UPC and the Masdar Initiative' explained another member of the UPC. 'Many people working in the Masdar Initiative ended up working in the UPC and vice versa' he continued, pointing out that 'at the very top of Masdar City and the UPC are ultimately the same people: the sheikh and the royal family.' As a result of this crossover, Estidama mirrors the characteristics of the creature of the Masdar Initiative and, unsurprisingly, favours the Emirati eco-city project. Masdar City passed the UPC's test in early 2010, receiving two Pearls. What is also important to note is that, back then, only 15 per cent of the project had been implemented, and Abu Dhabi's alleged eco-city consisted mostly of a large and dusty construction site. Masdar City did not exist, apart from a few buildings used as research facilities. Nonetheless, for the local planning council, the new Emirati city met the requirements of Estidama, and the Masdarian case was quickly closed.

Since the beginning of the construction of Masdar City, the role of Abu Dhabi's planning council has been limited to what a representative from the UPC described in an interview as 'occasional conversations with the Masdar Initiative,' and to sporadic examinations of master plans. As Masdar City is a project in constant evolution, its master plan changes often and, when it does, UPC planners have to be informed. This means that the developers must send digital or paper copies of

the renderings of new buildings, infrastructures and districts to the planning council. Overall, the UPC's evaluation of the Emirati eco-city project has been based
solely upon planning documents provided by the Masdar Initiative. The planning
council has carried out no *in situ* examinations, and it possesses no empirical data
on the effective environmental impact of the new settlement. 'Our job ends when
the project is approved' clarified one UPC manager. 'We would not go on our
own initiative and monitor the Masdar development,' the manager stated,
remarking that the role of the UPC is neither to check the performance of Masdar
City nor to request information on its sustainability. There is thus an evident
contradiction between what theoretically the UPC is meant to do via Estidama,
and the actual regulation of Masdar City. In practise, apart from monitoring
drawings which represent an idealized and ultimately abstract version of Masdar
City, the local planning council does not control what the Masdar Initiative is
building in reality. To further aggravate the situation, the UPC is not the only
actor, part of the Masdarian enterprise, denying responsibility for the sustainability
of the new city.

The eco-city project of Abu Dhabi has received unconditional support from
WWF UAE, which is the Emirati branch of the World Wildlife Fund. WWF
UAE (2008, 2011: no page) has described Masdar City as the 'world's greenest
city,' assuring that it is 'a city that will not just preserve the existing regional biodiversity but enhance it,' thereby setting 'a global benchmark for sustainable urban
development.' However, despite these lofty claims, WWF UAE ignores the
environmental impact of the Masdarian development. As a prominent member of
WWF UAE admitted in an interview, the organization is not in possession of any
data on the environmental sustainability of the settlement. In order to expose
possible discrepancies between what WWF UAE was claiming and what it is was
actually doing, the interviewee was asked if its organization was checking the
ecological footprint and impact of the new city on the biodiversity of the region.
He immediately replied with a 'no' and then added:

> but I am sure that there has been some kind of verification, because our col-
> leagues from the Masdar Initiative are working with a lot of international
> organizations to ensure that Masdar City is sustainable, but from our side we
> have not been involved in that process because that kind of work is outside
> our scope.

Like in the case of the UPC, the support of WWF UAE is groundless. The Emirati
eco-city project is deemed sustainable by an environmental agency which has not
verified the sustainability of the new settlement.

Furthermore, the nature of WWF UAE, similarly to that of the local planning
council, is intrinsically linked to the Masdar Initiative. Conventionally, WWF can
be defined as a non-governmental organization whose resources come from a mix
of private and public donations, but this description does not apply to WWF UAE.
In the United Arab Emirates, WWF can be funded exclusively by the local

government and it is bereft of the possibility of accepting private donations (which normally make up the major source of funding for WWF's national branches). The Masdar Initiative, as a public company of the government of Abu Dhabi, is one of the main funders of WWF UAE. 'We get most of our funding from Masdar' disclosed the same member of WWF UAE. The language that he used in the interview to refer to the Masdar Initiative is also emblematic. For this person, people working for the Masdar Initiative are *colleagues* or, in other words, associates who are part of the same institution. The institution in question, the Masdar Initiative, does not appear to care much about the sustainability of its experimental project, unless business matters are at stake. In ten years, the Masdar Initiative has produced only one environmental impact assessment. The task was commissioned to an advisory company named Hyder which released a short report in April 2009. The document (supposed to evaluate the environmental effects of the implementation and metabolism of the new city) offered no empirical evidence. Instead, it presented a series of hypothetical projections regarding the evolution of Masdar City. In three pages, Hyder speculates that Masdar City will probably not have a negative environmental impact, will not consume soil resources and will minimize waste. What Masdar City actually is and does, the document does not tell. In the summer of 2011, the Masdar Initiative removed Hyder's report from the official website of Masdar City, leaving no trace of it. As of this writing, nobody seems to be monitoring the environmental performance of the Emirati eco-city project.

In the case of the smart-city agenda of Hong Kong, the situation is more complex since, as elucidated in Chapter 5, smart urbanism in the SAR is the sum of many different initiatives taking place in different locations, under the agency of different makers. The privatization of a large share of the built environment of Hong Kong poses a serious challenge when it comes to monitoring the sustainability of smart interventions. There is a tension between what the SAR's Planning Department wants to achieve in theory and what it can do (and does) in practise. Replying to a question asking whether the concept of sustainability was part of the smart-city agenda of Hong Kong, a senior planner from the Planning Department commented with a straight 'sure.' 'In our last report,' she said, 'sustainable development is put as one of our overarching planning goals.' However, when the senior planner was asked about what the Planning Department is actually doing to ensure that Hong Kong's smart-city initiatives are sustainable, her reply was: 'It is very complicated. You see, most of the buildings are buildings owned by private companies or individuals, so we can't really force them to do anything. We are simply offering suggestions. Nothing more than that.' Only words and no action. The Planning Department talks about sustainability and the word *sustainability* itself is mentioned in many of its reports and recommendations, but this is a word that is not put into practical effect.

During the same interview, the senior planner discussed the lack of control that characterizes the SAR's urban areas currently on a lease, where private developers employ private companies like IBM to turn their properties into smart buildings. 'We have not done any sustainability assessment' she revealed. 'We simply cannot. If we tried, the owners of the buildings would object' the planner further explained, pointing out that the hands of the Planning Department are tied. 'We have to respect property rights' was her conclusion. Out of the jurisdiction of the Planning Department, the built environment grows uncontrolled, evolving in a way meant to please its owner. Particularly from an environmental sustainability perspective, the deregulation of the smart interventions led by the private sector, means that the sole local administrative division with an expertise in environmental protection and ecology, the Environment Bureau, has no chance to intervene. A series of interviews with representatives from the Environment Bureau, confirmed that the government's ecologists are not allowed to check the smart-city initiatives carried out by private developers. Simply put, in the majority of the SAR (the dark grey areas in Figure 5.1), no one can monitor the ecological impact of smart urbanism.

The situation does not change when *smart* takes place in those few urban areas that have not yet been privatized. As emerged in an interview with two policy-makers responsible for *Hong Kong Smart City Blueprint*, the Environment Bureau does not participate in the state-led smart interventions: its expertise has not been requested by the government. Although legally it would not encounter any barrier from the private sector, the Hong Kong Government has chosen not to include the Environment Bureau (and therefore the science of ecology) in the imple-mentation of the Central Business District Two and the Hong Kong Science and Technology Park. 'The Environment Bureau is not involved in the construction of the Park,' confirmed one of the directors of the Science Park in an interview. Since 2005, the Park has been growing with the addition of new buildings and facilities. Over the years, different construction and engineering companies such as AECOM (2008) have been employed by the government to increase the urban mass of the project. No one has the responsibility of monitoring its impact on the bioregion.

Conclusions

The unsustainability of Frankenstein's experiment was not inevitable. After all, the Doctor did manage to create life. His long experimental project led to the pro-duction of something truly astonishing: a human being capable of action and thought. Alas, the creature is a monster and its birth signs the beginning of the end for Victor and for those whom the young scientist loves the most. Shelley's *Fran-kenstein* is indeed a tragedy, but the story could have taken a different and more positive turn if the experiment had been conducted differently. Similarly, the unsustainability of the two experimental urban projects discussed in this book was not unavoidable. The Masdar City project and the smart-city agenda of Hong Kong have brought into existence an impressive array of urban spaces. The Emirati eco-city initiative has given birth to a city built from scratch in the harsh desert climate of Abu Dhabi, while the SAR's smart urbanism has led to the rise of new buildings and districts animated by technologies which were unthinkable until recently. Both urban experiments are unsustainable especially from an environ-mental and social point of view and, yet, their lacunas could have been anticipated.

Frankenstein's creature gradually became monstrous: it was not born a monster and, above all, it was not imagined as a monster by his maker. The experiment was supposed to produce the paragon of beauty and sustainability. This final section revisits the key three reasons why the creature of Victor Frankenstein and as such the creatures of the Masdar Initiative and Hong Kong's smart-city programme turned into monsters, suggesting what the protagonists could have done to avoid a tragic finale.

The first problem, due to the fragmented shape of the experiment and the incompatibility of some of its components, is essentially a planning issue. In terms of sustainability and design, Masdar City and Hong Kong's smart-city initiatives fail to be homogeneous urban spaces, because they grow in an uncoordinated manner. A fundamental assumption in experimental urbanism is that there is a scientific approach to urban development based on a holistic and rigorous plan of action (Cugurullo, 2018; Evans et al., 2016; Evans and Karvonen, 2011). At the core of the scientific method is the orderly arrangement of steps, such as calculations, observations and reflections, to accomplish an end. When it comes to urban experiments, the nature and order of these steps is, in theory, determined by a master plan which defines what has to be built, how, where and when. In practise, however, experimental urban projects can be very disorganized endeavours. More specifically, when eco-city projects and smart-city initiatives operate on a small scale, producing a single building, infrastructure or technology, the act of experimenting is regularly disciplined by a scientific method. A smart building, a smart grid or a roof-mounted solar panel, for example, is developed and implemented by systematically following complex engineering and computer science methodologies. At this scale, the process through which an abstract idea becomes a concrete artefact is disciplined. This, of course, does not imply that the process is error-free. The supposed advantage is the presence of a vision and a strategy which should prevent unexpected outcomes. At a larger scale, when the development of a whole city or an entire region is taken into account, experimental urban projects are often not framed by a comprehensive master plan: they evolve without overarching directions, strategies and limits in a chaotic way, thus developing heterogeneous shapes and, problematically, growing core elements that turn out to be in contrast with each other.

By exposing the planning process behind the genesis and development of Masdar City and *Hong Kong Smart City Blueprint*, the book has shown that heterogeneity and chaos can be found in the foundations of urban experiments. Such a condition of disorder, however, intended as a chronic lack of regular and methodical arrangement, is typical of cities and goes beyond the sphere of urban experimentation. Urban historians like Mumford (1961) and Benevolo (1993), for example, have repeatedly stressed that, historically, cities have rarely been uniform artefacts shaped by a homogeneous vision. Instead, they have grown through what were, in most cases, uncoordinated and fragmented processes of urbanization. For Mumford (1961), urban settlements, ranging from small towns to imperial capitals, have been the product of intersecting visions cultivated and subsequently advanced by diverse groups of people bearing different interests. It is this diversity and clash of urban visions that has resulted in built environments characterized by heterogeneous and often incongruous elements. In the Middle Ages, for instance, as Benevolo (1993) points out, the city did not have a single centre: it had several centres reflecting the agendas and the ideas of the then key actors. 'The physical form of the city depended upon the political organisation' of the city itself which 'served as a stage for the meetings and conflicts of many players' (Benevolo, 1993: 40). The experimental cities examined earlier resonate therefore with a historical tension in urban development. There is a deep and long-standing problem with the city imagined in abstract terms as an artefact which can be tidily arranged and moulded, in contrast with real cities which, as hyper complex social systems, grow under the pressure of multiple actors, in a way that constantly resists uniform and rigid urban designs.

The above urban condition is one crossing time and space. Far from the European cities discussed by Benevolo, in sixteenth-century Mexico, while he was exploring the New World with a band of conquistadores, Cortés (2004) vividly described in a letter to his king, the capital of the Aztec empire, Tenochtitlan, as a patchwork of diverse buildings and neighbourhoods controlled by different political forces. Similar examples abound later in the Baroque Era, during the Enlightenment and, with the birth of the industrial city, in the late modern period (Conforti, 2005; Mumford, 1961). Rigid master plans were rarely successful. Projects for ideal cities, like those mentioned in Chapter 1, all failed because it is extremely hard, if not impossible, to unify all the different visions, needs, interests and desires of the people who live, or will live in a given city, into a single urban design. Moreover, even if it was possible to create cities whose shape, structure and functions were disciplined by one overarching master plan, such an achievement might not necessarily lead to a desirable urban future. The echo of the warnings of the many scholars against the grandiose urban utopias of the past is still loud and impossible to ignore. Master-planned settlements established in a top-down manner can result in 'paternalistic' societies which would work 'only if you were docile and had no plans of your own,' argues Jacobs (1993: 24–25) claiming that 'as in all Utopias, the right to have plans of any significance' belongs exclusively to planners who easily forget the social diversity that makes cities truly alive. Likewise, Sennett (1970: 190) maintains that disorder, rather than order, should be an essential ingredient in urban development, for the actualization of 'personal styles and deviations' intrinsic to human nature, to be openly expressed, instead of being repressed by a purifying grand design (see, also, Rowe and Koetter, 1983; Wilson, 1992).

The above concerns are particularly valid in relation to the social sustainability of cities, but they are inadequate to process the severe environmental impact of urbanization. The case studies analyzed in this book clearly indicate that, without the presence of a precise and holistic master plan, experimental cities end up cultivating incompatible spaces, technologies and infrastructures which hinder each other's development, ultimately penalizing the overall environmental performance of the settlement. The cause of the formation of urban projects made of parts that do not act in concert is a planning void. Similarly, it is a lack of planning that is producing built environments which are not in sync with natural environments. Cities in balance with ecosystems cannot be born out of chaos and disorder: this is an urban vision calling for a rigorous plan of action which systematically takes into account and applies the science of ecology in the field of urban development. What is important to recognize is that planning a city is not necessarily an act of extremism and, as such, it does not require an extreme position akin to that of those scholars who are against master planning as a synonym for urban utopia. Master-planned urbanism and utopianism have historically often coincided, but this is precisely a coincidence and not an unavoidable phenomenon. There can be different degrees of master planning capable of steering urbanization away from socially monochromatic societies, in favour of places akin to those envisioned by urbanists such as Jacobs (1993) and Sennett (1970). The theories of ecological urbanism, discussed in Chapter 2, exemplify an understanding of planning as a tool for ensuring social diversity in the city. The work of Register (1987) and Krier (1998), in particular, pictures cities which are planned to be both ecologically and socially diverse. Conceptually, the challenge is then to imagine a type of master planning which, while being systematic, rigorous and comprehensive, does not fall into the trap of urban utopias, remaining inclusive, participatory and, above all, flexible enough to change according to the mutable needs of society and nature. Undoubtedly, there would be enormous practical challenges to implement a planning system of this kind and, yet, their scale is little compared to the current impasse in the imagination of alternative and more sustainable planning processes. This is an imaginative impasse that can be overcome only by thinking outside the box; the box in this case being the master planning that is conventionally associated with authoritarian urban utopias.

The second problem behind the unsustainability of the two case studies, is due to how the concept of urban sustainability has been formulated. Both the Masdar City project and the smart-city agenda of Hong Kong are underpinned by a very simplistic idea of what is sustainable and what should be sustained. The actual performance of these urban experiments is eventually unsustainable (particularly in social and environmental terms) because important notions of social and environmental sustainability have been disregarded by their makers. Fixing this problem, therefore, implies the formulation of more comprehensive urban equations meant to process the complexity of sustainability issues, rather than to focus on economic matters. Discussing the full spectrum of themes and questions of urban sustainability is outside the scope of this book, and the intention here is not to recommend a one-size-fits-all urban equation supposed to be applicable in every geographical location. This approach would be highly problematic since, as noted in recent studies, urban sustainability agendas should be tailored around specific contexts and open to bottom-up inputs: universality is not the answer (Caprotti et al., 2017). Instead, this section attempts to extrapolate from individual case studies, general lessons and reflections about what kinds of elements are needed for sustainable urban equations without presenting a definitive and universal list. There is a two-fold lacuna which compromises the formulation of Abu Dhabi's eco-city project and Hong Kong's smart urbanism, and in turn this can serve as a steppingstone to identify the two main categories of elements that require urgent attention from urban developers.

On the one hand, there are themes and issues of urban sustainability which are taken into account, but in a superficial and uncritical way. In this sense, the urban equations underpinning the development of Masdar City as well as Hong Kong's smart-city initiatives present elements which can potentially lead to the formation of sustainable cities. However, these elements are approached by the developers in a shallow manner and, thus, deeper problems of sustainability remain hidden during the growth of the city. An example of this lacuna is the issue of water. The Emirati eco-city project and numerous intelligent buildings in Hong Kong incorporate concerns about water, specifically in terms of *water waste*. There is a solid consensus with regards to water as one of the hardest and most pressing sustainability challenges of the century (Menga and Swyngedouw, 2018; Sivapalan et al., 2014). At first glance then, the environmental contribution that the Masdar City project and Hong Kong's smart urbanism achieve, by acknowledging the problem of water and tackling it through high-tech waste-management systems, is laudable. This is indeed a positive environmental outcome, but it is far from being sufficient and also hides a bigger problem.

Cities consume large quantities of water indirectly by simply growing. These flows of water are invisible and can be best understood by referring to the concept of *virtual water*. As Allan (2011) observes, water is directly consumed as a resource which people use, for instance, as a drink, but it can also be indirectly consumed by consuming products whose production requires water. Virtual water is, therefore, the invisible quantity of water embedded in the production of a product (Allan, 2003). Urbanization drains a lot of virtual water. It has been estimated that producing a single brick costs 2.02 litres of water, and that the virtual water of 1 m^2 of urban space amounts to 26,500 litres (Skouteris et al., 2018). In 2013 alone, urbanization in China produced 5.34 billion km^2 of urban space (Han et al., 2016). This is equal to approximately 140 trillion litres of water. Such water is invisible and yet real. It was consumed to expand existing cities and build new ones. Addressing the problem of water waste in cities is not enough. Unless the crucial issue of virtual water in urban development is integrated into current urban equations, environmental sustainability will always be out of reach.

On the other hand, there are significant urban sustainability issues which are not considered at all in the genesis and development of urban experiments. Referring back to the allegory of urban equations, such unconsidered issues are missing elements in the formulation of experimental urban projects: themes, ideas and concerns that are literally out of the equation. The urban equations at the core of Masdar City and *Hong Kong Smart City Blueprint* are simplistic, because several elements necessary for the maturation of sustainable cities have been completely ignored by the developers. The list is long with some of the critical gaps having been already mentioned earlier in this chapter and in the second part of the book. It was underlined that from an environmental sustainability point of view, themes like ecosystems and ecosystem services do not figure in the Masdarian formula and in Hong Kong's smart urbanism. From a social sustainability perspective, the provision of adequate housing is a question that has not been processed in the making of the smart-city initiatives of Hong Kong, while the notion of *community* is extraneous to the Emirati eco-city project whose implementation has resulted in the formation of a non-place.

Thinking in terms of urban equations, using the cases of Masdar City and Hong Kong as a starting point, can be a useful exercise to determine what other vital elements experimental cities are missing. *Happiness*, for example, is a theme which is lacking in both the Masdarian urbanism and in *Hong Kong Smart City Blueprint*. Urban and geographical studies indicate that, today, dense cities tend to cultivate states of malaise, sadness, stress and anxiety, particularly when access and proximity to green spaces are not part of the equation (MacKerron and Mourato, 2013; Okulicz-Kozaryn and Mazelis, 2018; Wyles et al., 2019). In spite of this dangerous problem, questions of happiness are not incorporated into the experimental urbanism of Abu Dhabi and Hong Kong, which is insensitive to how levels of wellbeing correlate with levels of urban density. *Sleep* is another case in point, a theme and a concern which, although crucial for human health, is often left out of the equation. Strong correlations exist between the quality and quantity of sleep and the design and metabolism of the city. While neuroscientists remark upon the necessity of uninterrupted sleep for mental and physical health, urbanists lament the proliferation of noise pollution and light pollution as defining features of cities since the Industrial Revolution (Crary, 2014; Gandy, 2017; Hong et al., 2019; Walker, 2017). *Sleep* could and should be a core element of any urban equation attempting to achieve urban sustainability. Like *ecosystem services, housing, happiness, community, wellbeing* and *sleep*, a plethora of other urgent themes are not being processed in the making of experimental cities which are consequently becoming deaf and blind to them. Following this line of thought, urban experiments formulated *à la* Masdar City and Hong Kong's smart-city agenda will ultimately devour resources from other cities and countries, unless issues of *supply chains* and *urban metabolism* are taken into account by their makers (Cugurullo, 2016; Murphy and Carmody, 2019; Rizzo, 2019). These experiments, like Frankenstein's monster, will also kill people. Not directly, because a city, unlike Victor's creature, does not have hands with which it can break a neck, but rather in indirect and arguably no less horrendous ways. If themes such as *education, justice, gender equality, public health, accessibility, disability* and *eldercare* are not engrained in experimental urbanism, the resulting urban experiments will be cities where a number of people will struggle to survive and self-realization will be impossible for many.

The third interconnected problem relates to a major flaw in the regulation of the built environment. The Masdar City project and Hong Kong's smart-city initiatives are producing severely unregulated spaces. Sometimes, as in the case of Estidama, a form of regulation does exist in the shape of a regulatory system meant to guarantee urban sustainability, but its effective control over what is being actually built is little or none. Often, due to the privatization of large portions of the city, controlling the evolution of experimental urban projects and their ecological impact, exceeds the remit of public environmental agencies. In essence, when urban experiments turn into monsters and start to devour and kill, no one is responsible for their actions. More problematically, such sheer lack of regulation is deeply connected to local political systems and, thus, it is particularly resilient to change and difficult to eradicate. As Imrie and Street (2009, 2011) observe, the regulation of the production of urban spaces is frequently a direct manifestation of local and national politics, with regulatory systems, codes and rules crafted *ad hoc* to protect and reproduce the dominant political system. This interpretation of urban regulations as systemic functions of the state, resonates with the cases of Abu Dhabi and Hong Kong. Estidama is a regulatory system created by the Emirati government to protect an investment of the government itself, while the deregulation of smart interventions in Hong Kong is a reflection of the SAR's neoliberal political economy. This is then a problem which cannot be fixed unless broader and harder political problems that are intrinsically part of the structure of the state are resolved first.

Taken together, these three problems undermine the validity of experimental urbanism, showing how easily this strand of urban development can fall short of addressing the unsustainability of cities. Cutting across the cases of Masdar City and Hong Kong and their tripartite failure is the same lacuna: urban experimentation is not driving real change. This finding resonates with the concerns exposed in Chapter 1, specifically about the doubtful sustainability of experimental urbanism. Numerous scholars have questioned the extent to which urban experiments are actually advancing alternative and more sustainable strategies of city-making (Castán Broto and Bulkeley, 2013; Cugurullo, 2016; Kaika, 2017; Karvonen et al., 2014; McGuirk et al., 2014; Savini and Bertolini, 2019). In relation to Abu Dhabi's eco-city project and Hong Kong's smart-city agenda, the answer is overall negative. *Change* is happening, but only in the sphere of technology. There is a significant degree of experimentation in both Hong Kong and Masdar City. The two case studies are remarkably similar in the way the city is employed as a laboratory to test and showcase cutting-edge machinery. However, in these eco and smart-city initiatives, experimentation is fixated on producing novel technologies. Planning processes, ideals of sustainable urbanism and regulations of the built environment, are far from being novel and alternative. On the contrary, they replicate long-standing and unsustainable dynamics of urbanization. In this sense, issues of fragmented planning schemes, market-driven ideas of sustainability and deregulated urban spaces are not symptomatic of experimental urbanism alone, and correspond with the same flaws that have been repeatedly attributed to urban projects developed under the banner of sustainability (Imrie and Lees, 2014; Krueger and Gibbs, 2007; Lawton, 2018; Miller and Mössner, 2020; Mössner and Miller, 2015, Raco, 2005; Rosol et al., 2017).

In light of recent developments in the field of AI, the issue of (de)regulation might be the most pressing one. Frankenstein's creature was left alone, unsupervised and free to live by itself. It became independent or, in other words *autonomous*, because of the power to act on its own. As this book has shown in Chapters 4 and 5, part of the management of Masdar City and of Hong Kong's smart-city initiatives is showing a stepping back of their makers. Eco and smart-city developers are installing a wide range of automated and autonomous technologies, ranging from driverless cars to city brains, whose operation requires little or no input from human agents. This phenomenon can be understood as the passage from automation to *autonomy*, and its manifestation coincides with the growing emergence of artificial intelligence in the operation of cities. In a context in which novel AI urban technologies like Alibaba's City Brain could design their own urban equations without human supervision, it is not the experiment to be left alone in the dark. In Shelley's novel, Victor himself, the maker, is alone in the dark. The darkness of ignorance, as the Doctor ignores what the creature is thinking and what drives its catastrophic actions. He is also alone in the dark in his being impotent against the powerful monster that he has created. He cannot stop it. His strength is just not enough, and when Victor commits his life to end the life of the monster it is too late. The next chapter turns the narrative to the question of artificial intelligence in the development of experimental cities, tracking the path taken by Frankenstein's creature and trying to anticipate its next move.

References

Allan, J. A. (2003). Virtual water-the water, food, and trade nexus. Useful concept or misleading metaphor?. *Water international*, 28 (1), 106–113.

Allan, T. (2011). *Virtual water: Tackling the threat to our planet's most precious resource.* IB Tauris, London

AECOM (2008). Hong Kong Science Park. [Online] Available: https://www.aecom.com/projects/hong-kong-science-park// [Accessed 10 November 2020].

Batty, M. (2018). *Inventing future cities.* MIT Press, Cambridge.

Benevolo, L. (1993). *The European city.* Blackwell, Oxford.

Caprotti, F., Cowley, R., Datta, A., Broto, V. C., Gao, E., Georgeson, L., Herrick, C., Odendaal, N. and Joss, S. (2017). The new urban agenda: Key opportunities and challenges for policy and practice. *Urban research & practice*, 10 (3), 367–378.

Castán Broto, V., and Bulkeley, H. (2013). A survey of urban climate change experiments in 100 cities. *Global Environmental Change*, 23 (1), 92–102.

Conforti, C. (2005). *La citta' del tardo Rinascimento.* Laterza, Roma.

Cortés, H. (2004). *Five letters 1519–1596* (Vol. 10). Routledge, London.

Crary, J. (2014). *24/7: Late capitalism and the ends of sleep.* Verso, London.

Cugurullo, F. (2013). The business of utopia: Estidama and the road to the sustainable city. *Utopian Studies*, 24 (1), 66–88.

Cugurullo, F. (2016). Urban eco-modernisation and the policy context of new eco-city projects: Where Masdar City fails and why. *Urban Studies*, 53 (11), 2417–2433.

Cugurullo, F. (2018). Exposing smart cities and eco-cities: Frankenstein urbanism and the sustainability challenges of the experimental city. *Environment and Planning A: Economy and Space*, 50 (1), 73–92.

Elmqvist, T., Andersson, E., Frantzeskaki, N., McPhearson, T., Olsson, P., Gaffney, O., Takeuchi, K. and Folke, C. (2019). Sustainability and resilience for transformation in the urban century. *Nature Sustainability*, 2 (4), 267–273.

Evans, J. and Karvonen, A. (2011). Living laboratories for sustainability: exploring the politics and epistemology of urban transition. In Bulkeley, H.Castán Broto, V., Hodson, M. and Marvin, S. (Eds.). *Cities and low carbon transitions*, Routledge, London, pp. 126–141.

Evans, J., Karvonen, A. and Raven, R. (Eds.). (2016). *The experimental city*. Routledge, London.

Gandy, M. (2017). Negative luminescence. *Annals of the American Association of Geographers*, 107 (5), 1090–1107.

Gray, J. (2008). *Black mass: Apocalyptic religion and the death of utopia*. Penguin Books, London.

Han, M. Y., Chen, G. Q., Meng, J., Wu, X. D., Alsaedi, A. and Ahmad, B. (2016). Virtual water accounting for a building construction engineering project with nine sub-projects: a case in E-town, Beijing. *Journal of Cleaner Production*, 112, 4691–4700.

Hong, A., Kim, B. and Widener, M. (2019). Noise and the City: Leveraging crowd-sourced big data to examine the spatio-temporal relationship between urban development and noise annoyance. *Environment and Planning B: Urban Analytics and City Science*, 2399808318821112.

Imrie, R. and Lees, L. (Eds.) (2014). *Sustainable London?: The future of a global city*. Policy Press, Bristol.

Imrie, R. and Street, E. (2009). Regulating design: The practices of architecture, governance and control. *Urban Studies*, 46 (12), 2507–2518.

Imrie, R. and Street, E. (2011). *Architectural design and regulation*. John Wiley & Sons, London.

Jacobs, J. (1993). *The death and life of Great American cities*. The Modern Library, New York.

James, P. (2014). *Urban sustainability in theory and practice: circles of sustainability*. Routledge, London.

Kaika, M. (2005). *City of flows: Modernity, nature, and the city*. Routledge, London.

Kaika, M. (2017). 'Don't call me resilient again!': the New Urban Agenda as immunology… or… what happens when communities refuse to be vaccinated with 'smart cities' and indicators. *Environment and Urbanization*, 29 (1), 89–102.

Karvonen, A., Evans, J. and van Heur, B. (2014). The politics of urban experiments: Radical change or business as usual? In Marvin, S. and Hodson, M. (Eds.) *After Sustainable Cities*. Routledge, London, 105–114.

Keil, R. (2005). Progress report – urban political ecology. *Urban Geography*, 26 (7), 640–651.

Krier, L. (1998). *Architecture: Choice or fate*. Papadakis Publisher, Singapore.

Krueger, R. and Gibbs, D. (Eds.). (2007). *The sustainable development paradox: Urban political economy in the United States and Europe*. Guilford Press, New York.

Lawton, P. (2018). Uneven development, suburban futures and the urban region: The case of an Irish 'sustainable new town.' *European Urban and Regional Studies*, 25 (2), 140–154.

Long, J. and Rice, J. L. (2019). From sustainable urbanism to climate urbanism. *Urban Studies*, 56 (5), 992–1008.

MacKerron, G. and Mourato, S. (2013). Happiness is greater in natural environments. *Global Environmental Change*, 23 (5), 992–1000.

McGuirk, P., Dowling, R. and Bulkeley, H. (2014). Repositioning urban governments? Energy efficiency and Australia's changing climate and energy governance regimes. *Urban Studies*, 51 (13), 2717–2734.

Menga, F. and Swyngedouw, E. (Eds.). (2018). *Water, technology and the nation-state*. Routledge, London.

Miller, B. and Mössner, S. (2020). Urban sustainability and counter-sustainability: Spatial contradictions and conflicts in policy and governance in the Freiburg and Calgary metropolitan regions. *Urban Studies*. doi:0042098020919280.

Mössner, S. and Miller, B. (2015). Sustainability in one place? Dilemmas of sustainability governance in the Freiburg metropolitan region. *Regions Magazine*, 300 (1), 18–20.

Mumford, L. (1961). *The city in history: Its origins, its transformations, and its prospects*. Harcourt, Brace & World, New York.

Murphy, J. T. and Carmody, P. R. (2019). Generative urbanization in Africa? A socio-technical systems view of Tanzania's urban transition. *Urban Geography*, 40 (1), 128–157.

Okulicz-Kozaryn, A. and Mazelis, J. M. (2018). Urbanism and happiness: A test of Wirth's theory of urban life. *Urban Studies*, 55 (2), 349–364.

Raco, M. (2005). Sustainable Development, Rolled-out Neoliberalism and Sustainable Communities. *Antipode*, 37 (2), 324–347.

Register, R. (1987). *Ecocity Berkeley: building cities for a healthy future*. North Atlantic Books, Berkeley.

Rizzo, A. (2019). Predatory cities: Unravelling the consequences of resource-predatory projects in the global South. *Urban Geography*, 40 (1), 1–15.

Robertson, M. (2017). *Sustainability principles and practice*. Routledge, London.

Rosol, M., Béal, V. and Mössner, S. (2017). Greenest cities? The (post-) politics of new urban environmental regimes. *Environment and Planning A: Economy and Space*, 49 (8), 1710–1718.

Rowe, C. and Koetter, F. (1983). *Collage city*. MIT Press, Cambridge.

Savini, F. and Bertolini, L. (2019). Urban experimentation as a politics of niches. *Environment and Planning A: Economy and Space*, 0308518X19826085.

Sennett, R. (1970). *The uses of disorder: Personal identity and city life*. W.W. Norton & Company, New York and London

Shelley, M. (2013). *Frankenstein: Or, the Modern Prometheus*. Penguin, London.

Sivapalan, M., Konar, M., Srinivasan, V., Chhatre, A., Wutich, A., Scott, C. A., Wescoat, J.L. and Rodríguez-Iturbe, I. (2014). Socio-hydrology: Use-inspired water sustainability science for the Anthropocene. *Earth's Future*, 2 (4), 225–230.

Skouteris, G., Ouki, S., Foo, D., Saroj, D., Altini, M., Melidis, P., Cowley, B., Ells, G., Palmer, S. and O'Dell, S. (2018). Water footprint and water pinch analysis techniques for sustainable water management in the brick-manufacturing industry. *Journal of cleaner production*, 172, 786–794.

Swyngedouw, E. (2010). Apocalypse forever?. *Theory, culture & society*, 27 (2–3),213–232.

UPC (2015) Estidama. [Online] Available: http://estidama.upc.gov.ae/estidama-and-pearl-rating-system.aspx?lang=en-US [Accessed 10 November 2020].

Verma, P. and Raghubanshi, A. S. (2018). Urban sustainability indicators: Challenges and opportunities. *Ecological indicators*, 93, 282–291.

Walker, M. (2017). *Why we sleep: The new science of sleep and dreams*. Penguin, London.

Wilson, E. (1992). *The sphinx in the city: Urban life, the control of disorder, and women*. University of California Press, Berkeley.

Wyles, K. J., White, M. P., Hattam, C., Pahl, S., King, H. and Austen, M. (2019). Are some natural environments more psychologically beneficial than others? The importance of type and quality on connectedness to nature and psychological restoration. *Environment and Behavior*, 51 (2), 111–143.

WWF UAE (2008). WWF, Abu Dhabi unveil plans for sustainable city. [Online] Available: http://wwf.panda.org/index.cfm?uNewsID=121361 [Accessed 10 November 2020].

WWF UAE (2011). Our projects. [Online] Available: http://uae.panda.org/what_we_do/projects2/2f/ [Accessed 10 November 2020].

Yan, Y., Wang, C., Quan, Y., Wu, G. and Zhao, J. (2018). Urban sustainable development efficiency towards the balance between nature and human well-being: Connotation, measurement, and assessment. *Journal of Cleaner Production*, 178, 67–75.

7

ARTIFICIAL INTELLIGENCE AND THE RISE OF THE AUTONOMOUS CITY

Introduction

Artificial intelligence (AI) is one of the most overt enigmas of the twenty-first century. On the one hand, AI technologies are a familiar feature of contemporary societies, and their application can be found in a variety of fields (Yigitcanlar and Cugurullo, 2020; Stone et al., 2016). AI is present in medicine, in invisible software providing a diagnosis and recommending a therapy, or as a robotic arm performing a surgery (He et al., 2019). AI is in law, as an algorithm reaching verdicts which will determine who will go to prison and who is eligible for insurance coverage (O'Neil, 2016; Richardson et al., 2019). AI is in war, far away in the battlefield, embodied in autonomous weapon systems (Horowitz, 2019). AI is nearby, in the heart of the city where self-driving cars operate in real-life environments (Cugurullo et al., 2020; Milakis et al., 2017). On the other hand, however, it is unclear what an artificial intelligence actually is and does. There is no established universal definition of AI. Just as the concept of human intelligence is hard to define in precise terms, so the meaning of artificial intelligence is elusive. Just as human intelligence can reside in many different bodies, so artificial intelligence can have a plethora of heterogeneous incarnations. AI is obvious and yet enigmatic.

The aim of this chapter is to explore the meaning and practise of artificial intelligence in cities. There is a strong connection between AI and the city. AI technologies are developed, tested and implemented primarily in cities. Many of them are designed to function in the built environment and can be understood as *urban artificial intelligences*: artificial intelligences which acquire meaning and purpose from the city, while changing the meaning and purpose of the city itself (Cugurullo, 2020). By being integrated into urban spaces and infrastructures, these urban AIs impact on the city, altering its form and metabolism. At the same time, cities change AI. The city gives a place and a role to different AIs, thereby defining their identity. Moreover, the city gives AI materiality. In its purest form, artificial intelligence is intangible software: a soul without a body. AI finds in cities numerous bodies. From an autonomous car to a robot, and from a building to a whole city. AI is embodied in diverse elements of the urban fabric, thus becoming complete and fully operational. In addition, through this process of incarnation, the city exposes the contradictions, problems and dangers of AI. It is only when an AI is operating in a real-life environment that theoretical problems become evident, and also concrete. Part of this chapter is devoted to identifying and discussing the most critical issues caused by the implementation and diffusion of urban artificial intelligences.

The structure of this chapter is tripartite. First, the chapter clarifies the meaning of *artificial intelligence*. Drawing mostly upon philosophy and computer science, the next section elucidates the key conceptual and technological characteristics of artificially intelligent entities. Frankenstein's creature is employed as an analogy to explain how an AI can acquire knowledge and then act according to what it believes is the right thing to do. Here Shelley's tragic character also becomes an entry point to debating controversial aspects of AI. In particular, the book questions the ability of artificial intelligences to engage with ambivalent concepts like *good* and *bad*, and it raises concerns over the fact that the goals of an AI might not necessarily be aligned with human goals. Second, the chapter examines the main manifestations of AI in cities. The growing presence of artificial intelligences in the built environment is not interpreted as an abrupt phenomenon, but rather as a long-term process of technological and urban development, culminating in the passage from *automation* to *autonomy*. This passage is illustrated by unpacking the difference between Masdar City's automated and autonomous urban transport systems. Subsequently, the narrative shifts to self-driving cars, robots and city brains, which are presented as the principal categories of urban artificial intelligence. The discussion focuses on these categories, revealing the diverse roles, functions and materialities which AI can possess in the city. Third, after having delineated the contours of what the book calls the *autonomous city*, the chapter delves into the sustainability implications of the diffusion of urban artificial intelligences. As a disruptive technology, AI is shaking the social, environmental, political and economic dimensions of the city, redefining the essence of the *urban* that was observed in Chapter 1 through the lens of Aristotle's philosophy. The penultimate section recalls the general theoretical and technological problems of artificial intelligence examined in the second section, and discusses them specifically in relation to the urban categories of AI covered in section three. In the end, the chapter finds in the constant tension between AI and ethics, the mother of all challenges which cities face in the era of artificial intelligence, arguing that without philosophical inquiry and intellectual debate technological innovation alone will not suffice to make the autonomous city a sustainable city.

Artificial intelligences

The concept of artificial intelligence eludes a single universal definition and its history is intricate (Cave et al., 2020; Clifton et al., 2020). However, building upon decades of studies on the topic, it is possible to define the main characteristics of an artificially intelligent entity. The first word composing the term *artificial intelligence* is relatively simpler to unpack. Something is normally regarded as being artificial when it does not exist in nature and does not arise from a natural process. On these terms, an artificial intelligence is not the outcome of, for instance, the million-year-old evolutionary process that led to the development of the human brain. Instead, it is something that can be human-made or, as Bostrom (2017) remarks, created by intelligent machines. As such, it is assumed that an artificial intelligence resides in an artefact. Many AIs, especially those found in cities, are embodied, meaning that they possess an otherwise inanimate artefact (such as a computer, a car or even a whole building) which becomes an intrinsic part of them (Cugurullo, 2020).

The second word, *intelligence*, has been the subject of many studies since the birth of philosophy. More recently, it has been at the core of the research agenda of modern disciplines, such as cognitive psychology and neuroscience, which have developed robust methodologies to empirically examine how humans manifest intelligence. Yet, the nature, source and location of human intelligence remain mysterious and debatable. Aristotle was one of the first philosophers to explore the notion of intelligent behaviour, and his theories are still applied in contemporary AI research. Earlier in the book, Aristotle's philosophy was employed to shed light on key aspects of the urban, and this section draws again on his work to identify the foundations of intelligence. A major contribution of Aristotelian philosophy is the concept of *syllogism* whose formulation and application can be seen as a manifestation of intelligence (Aristotle, 2009). A syllogism is the reaching of correct conclusions on the basis of correct premises which are, in turn, based on knowledge. A classic example of a syllogism is:

Socrates is a man (premise 1). *All men are mortal* (premise 2). *Hence, Socrates is mortal* (conclusion).

Centuries later, Descartes (1596–1650) investigated an interconnected question, reflecting on the location of intelligent behaviour. He argued that part of the human mind and, therefore, part of intelligence is immaterial (Descartes, 2003). The argument of the French philosopher and scientist was grounded in metaphysical concepts such as *soul* and *spirit*, and urged the scientific community to think of the human mind not simply as a machine, and of human intelligence as a non-mechanical operation. However, the argument of Descartes was challenged first by empiricist philosophers like Locke (1632–1704) and Hume (1711–1776), and subsequently by eminent members of the Vienna Circle: Wittgenstein (1889–1951), Russell (1872–1970) and Carnap (1891–1970). Common across these philosophers is the understanding of the mind as a physical system which acquires knowledge through the senses, and uses it to make decisions. Today, the field of AI research largely embraces empiricism and the philosophy of the Vienna Circle, positing that artefacts, such as computers, cars or even an entire city, for example, can acquire knowledge from sensory experience and make decisions. From an Aristotelian perspective, therefore, an artificial intelligence can be conceptualized as an artefact which has the capacity to acquire knowledge by sensing the environment, then using that knowledge to develop correct premises, and eventually building upon those correct premises to reach the correct conclusions.

In more technical terms, an advanced AI would feature a series of key elements and skills. First, an AI would have the ability to learn, intended as *gaining knowledge*. AIs normally gain knowledge by acquiring data either through the perception of the surrounding environment or via pre-existing data sets (Russell and Norvig, 2016). In the first case, the AI collects the data itself by means of, for instance, cameras and microphones. In the second case, large data sets, namely *big data*, are installed by the developers. Second, an AI would be capable of making sense of the data by extracting concepts from it (Bostrom, 2017). Perceiving the constant falling of a roughly spherical mass of water from clouds (a drop) can be associated with the concept of rain. The concomitant perception of a booming sound produced by the rapid expansion of atmospheric gases heated by lightning (a thunder) can be in turn associated with the concept of a thunderstorm. Third, an AI would be able to deal with complex situations in which some information is missing, or part of the collected data is unclear. This is, in essence, the ability of handling uncertainty (Kanal and Lemmer, 2014; Pearl, 2014). Fourth, an AI would make decisions and then act according to those decisions in a rational way. For AI scholars, *doing the right thing* and *acting rationally*, are about the achievement of the best possible outcome, where *best* is understood in relation to pre-designed performance measures which define what is *right* or *wrong* (Russell and Norvig, 2016). Drawing upon the Aristotelian syllogism mentioned earlier, an AI capable of perceiving Socrates, aware that all men are mortal and designed to protect all men, should not decide to kill Socrates. Killing Socrates would not be the right course of action because, in this example, the artificial intelligence is meant to support the life of men like Socrates and, thus, the protection of men becomes the key criteria against which its performance is assessed. In the case of an advanced AI, all these features and skills would be present and used autonomously or, as Levesque (2017: 3) concisely explains, in an *unsupervised* manner.

There are of course many grey areas in both the theory and practise of artificial intelligence, and even the simplest syllogism can hide enormous conceptual and technological challenges. Mary Shelley's *Frankenstein* offers several important parallels to critically reflect on how artificial intelligences function, with an emphasis on the most problematic aspects of AI technologies. The creature generated by Doctor Victor Frankenstein has, from the philosophical standpoint of the Vienna School, an artificial intelligence. It is gifted with the following capacities. It can sense the surrounding environment. It can acquire knowledge through the senses. It can use that knowledge to formulate decisions and act upon them. This intelligent behaviour is not the process of natural evolution. The creature is the result of a human-made experiment and is *de facto* an artefact assembled by the Doctor. When they are observed through the lens of empiricism, the mind and the intelligence of Victor's creation display a physical dimension. On these terms, the creature is not dissimilar from a machine with sensors which acquire information. The type of this information is diverse, since it is produced through a diverse sensory experience: the creature sees, touches, smells, tastes and hears, like humans do.

The sensorial abilities of Frankenstein's experiment are very well developed. The problems arise with the other aspects of its artificial intelligence. Although the creature can acquire data from the surrounding environment, it does not possess any pre-installed data set. From this point of view, the creature is a blank canvas. It also does not have anyone who, right after its genesis, helps it to acquire knowledge and, above all, contextualize that knowledge. As a result, the intelligence artificially created by the Doctor is not initially capable of extracting concepts out of the data that it is rapidly absorbing through sensory experiences. This is where the real tragedy begins. From the perspective of physics, for instance, the creature is not aware of its immense strength and of the results that the application of this strength on other bodies can have. It does not know that bones can break, and that the breaking of certain bones can kill a human. More in general, the creature has no concept of death. Going back to the Aristotelian syllogism, it cannot develop the second premise, *all men are mortal*. In addition, it does not have the capacity to deal with complex situations in which data is missing and, consequently, the outcomes of its actions are often catastrophic. Finally, because of its lack of pre-designed performance measures, it ignores what is right and what is wrong within the specific context where it acts. As a result, some of its actions, such as the killing of human beings, go against what is understood as the *right thing* at that time and in that place. The artificially intelligent creature has no knowledge of even the most basic social norms. Everybody around it knows that depriving people of their life is wrong unless the law says otherwise. It is clear to anyone that murdering a child is particularly wrong. Frankenstein's creature knows neither these norms nor the law, and its first murder is not an act of malice, but rather an act of ignorance. It becomes a monster without knowing it.

In this sense, Frankenstein's creature can be conceptualized as a *seed AI*. From a technical point of view, a seed AI is an artificial intelligence whose initial features and skills are limited in number, scope and power (Yudkowsky, 2001). This type of AI starts with little or no data. However, it has a strong potential connected with a strong learning capacity, meaning that the machine can absorb knowledge quickly, thereby increasing its intelligence at a fast pace. There is a conceptual thread linking the notion of a seed AI with Alan Turing's theory of the *child machine*, which illuminates some critical aspects of artificially intelligent entities. According to Turing (1950), instead of creating a programme which simulates the mind of an adult, it would be better to design one which simulates the mind of a child. For him, if such a programme was subsequently 'subjected to an appropriate course of education,' it would eventually obtain the intelligence of an adult (Turing, 1950: 456). This rationale is based first on the premise that it is easier to artificially create the mind of a child, compared to that of an adult. The former would be architecturally simpler inasmuch as, although most brain cells are formed before birth, the majority of the connections among them (the synapses) develop during childhood. Above all, being the intelligence of a child still in the making, it could be more malleable, than the intelligence of an adult with a crystallized set of knowledge and concepts. Applied in the context of AI research, the development of a seed AI would facilitate the alignment of the concepts that an AI has and uses to drive its actions with those of the developers, particularly with regards to key ideas such as *right* and *wrong*. In other words, seed AIs might be the easiest way to make sure that human beliefs and knowledge are shared by artificial intelligences.

However, while technologically easier, the development of a seed AI could be more complex both in practical and philosophical terms. The educator, or educators, of a seed AI would need to be constantly present, since this type of AI, as discussed above, learns very fast and can begin to act quickly. Victor Frankenstein abandons his creature right after he finishes the experiment. The creature is left alone to learn, and it develops concepts which differ substantially from those of its creator and, more generally, from basic social norms. When Victor establishes contact again with his creature, the ideas of the latter are already crystallized and its mind, like that of an adult, is now hard to reshape. Even if the Doctor had been constantly present during and after the experiment, educating his creature and preventing it from becoming a monster, there would still be broader pending questions. Philosophically, for instance, there is the question of who educates the educator and of the subjectivity that underpins concepts such as *right* and *wrong* which are far from being universal, and whose understanding and practise have been historically changing across time and space. This would not be a problem per se if the agency of the AI was limited to the boundaries of a laboratory, thus affecting only its creators, but this is not what artificial intelligences are normally made for. AI developers build AIs in laboratories, with the aim of eventually applying them in real-life environments. Frankenstein's creature was born in a lab, but its actions take place outside it, and impact on a number of people, not only on Victor. As noted in Chapter 1, this is also a common dynamic in urban experimentation with the experiment usually beginning in an urban area employed as a lab and later escalating.

Furthermore, there is the issue of the verticality or hierarchy of concepts, in relation to horizontal and non-hierarchical contexts. Assuming that an AI learns a set of concepts like *right, wrong, good* and *bad*, it might not be clear how such general and absolute concepts should be applied to variegated situations in which there are many shades of grey between black and white. In other words, doing the right thing is often an action taking place in a grey area. For example, there can be situations in which saving someone's life could be possible only by causing the death of another person. If an AI is educated in a way that the concept of killing a human is wrong and the concept of preserving the life of a human is good then, theoretically, this given AI should always prioritize the preservation of human life. However, in practise, in a hypothetical scenario in which not killing someone is equal to not preserving a human life, the original hierarchy of the concepts at the foundation of the AI gets severely undermined. The crux of the matter is what course of action would an AI educated to protect humans prioritize, when every possible action inevitably leads to the death of somebody.

This issue relates to the so-called *trolley problem*: a classic thought experiment in ethics, which nowadays has often been applied to cars driven autonomously by an artificial intelligence (Nyholm and Smids, 2016). One of the main promises of autonomous cars is that, by eliminating human errors, they can considerably reduce car accidents but, as noted by several scholars, this does not mean that there will not be accidents anymore (Bonnefon et al., 2016; Goodall, 2014; Lin, 2016). Accidents might become rarer, but the possibility of a car accident would still exist. Following this logic, the thought experiment involves imagining a situation in which harm is unavoidable. Perhaps, the brakes of the autonomous car are malfunctioning or, because of an unexpected factor (such as a human suddenly crossing the vehicle lane), the AI does not have enough time to stop the car. In this case, the question is how the AI will choose to distribute harm that is inevitable (Awad et al., 2018; Bonnefon et al., 2019). The AI could, for instance, decide to turn abruptly and crash the car into a wall thus killing its passenger, or it might stay on course and kill a pedestrian. The question and the thought experiment become more complex if, hypothetically, more than one pedestrian would be hit and killed. Alternatively, it may be that only one pedestrian would die, but she is a pregnant woman, or maybe the person in question is elderly or a criminal. The possibilities are numerous and, as current studies show, there is no universal ethical answer to what is right or wrong in such grey areas (Awad et al., 2020). Furthermore, even if the AI develops a clear understanding of the right thing to do, its following actions might still cause harm to humans. Frankenstein's creature itself is conscious that killing its creator is wrong and, yet, the story ends with the death of Victor.

It is important to note that, when it comes to AI, using *Frankenstein* as an analogy can also be misleading. Caution is needed because this is an analogy that risks anthropomorphizing artificial intelligence. There are three critical aspects to highlight. First, although monstrous, the result of Victor's experiment is essentially a man. As emphasized in the previous chapter, the Doctor wanted to create an ideal human being and, in order to do so, he had selected the best physical human elements that he could find. The creature is composed of human muscles, organs, bones, eyes and it has, therefore, a human structure and appearance. An artificial intelligence, however, can be located in non-anthropomorphic entities. This means that an AI does not necessarily look like or even remotely resemble a human being. Several examples prove this point. Autonomous cars and drones driven by an artificial intelligence, for instance, have a physical structure and an appearance strongly non-human. Both these AI technologies come in a variety of sizes and shapes, and none of them, given their purpose and application, can be designed in an anthropomorphic manner. Cars need wheels to move and an empty interior to accommodate passengers. Drones require wings and blades to fly, and their closest physiological connection with a biological creature would be with birds or insects. Second, there are artificial intelligences which are intangible. While incarnated in a computer hardware made of microchips, a solid case and power supply units, some AIs consist of pure software which implements a series of immaterial instructions or, using the language of computer science, *algorithms*. Bodiless AI software can be found in a plethora of domains ranging from healthcare (a basic AI uses speech recognition and cameras to examine a patient, check symptoms against a large data set, and prescribe medications) and customer service (an AI chatbot conducts a conversation with a client), to law (an AI rapidly reviews a case, checks it against a body of laws, and reaches a verdict) and education (an AI creates a virtual reality in which students immerse themselves in a given subject).

Third, Frankenstein's monster not only looks like a man, but it also feels like a man. Although its feelings and consequently its actions are often confused, the emotions of the creature are ultimately human in nature. It can feel love, hate, happiness and sadness, and its agenda is driven by typically human desires and motivations such as curiosity, vengeance and pity. An artificial intelligence, however, can be animated by feelings that completely escape the spectrum of human emotions. As Bostrom (2017) points out, the motivations of an artificial intelligence do not necessarily have to be human-like, inasmuch as an AI is a non-human entity (see also Yudkowsky, 2008). As a biological organism, Frankenstein's creature, for instance, needs food to sustain its life and, thus, its actions are partly driven by hunger and self-preservation. Similarly, albeit considerably stronger than an average human being, it is vulnerable, hence its plans to find shelter. In addition, the creature is not asexual: a condition which makes its longing for a partner and its aversion to a life of complete solitude, plausible. However, the same cannot be said for AI technologies which, as noted above, are structurally different from humans and can have no physical structure at all. In AI research, these tensions are unpacked in the *orthogonality thesis* which sees intelligence and motivations as two orthogonal axes moving along different directions (Armstrong, 2013; Bostrom, 2012). By separating the course of intelligence from the development of goals, this thesis posits that different intelligences can be driven by different motivations. On these terms, 'more or less any level of intelligence could in principle be combined with more or less any final goal,' meaning that the actions of an intelligent entity can be motivated by a wide range of possible goals (Bostrom, 2017: 130). Ultimately, according to the orthogonality thesis, an AI might have 'utterly non-anthropomorphic goals' (Bostrom, ibid.).

Cities and AIs

The city is the arena where AI technologies are being studied, tested and implemented (Allam and Dhunny, 2019; Batty, 2018). Artificial intelligence finds in the urban a number of incarnations, across different scales and domains. These urban incarnations of AI are referred to as *urban artificial intelligences*. They are artificial intelligences which are embodied in urban spaces, urban infrastructures and urban technologies (Cugurullo, 2020). As mentioned earlier, AI can exist as pure software and it does not necessitate a body. Urban artificial intelligence is a type of AI which requires a body to function. As this section will show, the material embodiment of urban artificial intelligence ranges from a car to the entire city. Because of their diverse materiality, urban artificial intelligences have multiple shapes, functions and purposes, but they all share the general traits of AI discussed in the previous section. Urban AIs are autonomous artefacts capable of gaining knowledge about the surrounding urban environment and making sense of the acquired data, using it to act rationally according to pre-defined goals in complex situations and spaces in which some information might be missing and, above all, humans are not steering their actions.

Urban artificial intelligences not only share the main characteristics and capabilities of AI, but they also share the same challenges, pressures, limits and concerns related to artificially intelligent entities. For this reason, the issues and questions that AI is raising, whether technological or philosophical in nature, are fundamentally urban issues and urban questions, and it is therefore essential to understand how AI manifests itself in cities and with what effects. These spatial matters must also be observed from a temporal perspective, since the manifestation of AI technologies in the built environment is not an abrupt phenomenon changing, all of a sudden, the metabolism and design of the city, but rather part of a long-term process of technological innovation which has been gradually affecting urban development. More specifically, the entrance of AI in cities is the outcome of the passage from *automation* to *autonomy* in the management of urban infrastructure and services. This is a key moment in the evolution of the city when, for the first time in history, an artificial intelligence gets behind the operation of elements of cities, which have been traditionally controlled by a human intelligence.

To understand this passage and the fundamental difference between *automation* and *autonomy*, the case of Masdar City offers important insights. Masdar City's Personal Rapid Transit (PRT), discussed in Chapters 4 and 6, exemplifies automation. The Masdarian automated transport system functions in the following way. In terms of infrastructure, the streets of Masdar City are filled with smart sensors defining the tracks over which the driverless vehicle will pass. Sensors are placed in a sequential manner, with a sequence of sensors forming a precise PRT track. Together, the tracks compose a well-defined transport network which is situated in the undercroft of the new city (see Figure 7.1). PRT vehicles cannot escape the undercroft, as they cannot leave the tracks marked by the sensors. Passengers can enter a PRT car only from a PRT station which looks and operates like a common metro station. Once inside, the passenger chooses the destination by means of a touchscreen. The options are limited to the PRT stations that have been built. As soon as one of the pre-defined destinations has been selected, the PRT machine starts to move towards it. The trip takes place in an automated way because PRT vehicles are programmed to follow a specific track of the grid. They do not choose the route, instead adhering to the patterns established by the planners of Masdar City. Each trip is the same with the machine responding, every time, to the call of the sensors inserted by a team of engineers in line with software designed by a group of

FIGURE 7.1 View from inside a PRT vehicle in motion. Visible on the ground are the tracks followed by the automated cars. The environment shown in this image is Masdar City's undercroft, a confined space which no other type of vehicle can traverse. Pedestrians are not allowed and barriers enclose the undercroft to prevent potential intruders from entering this space.

Source: Author's original

computer scientists. There is no room for variations or improvisation. The PRT is an automated technology because its actions are disciplined by *a priori* decisions made in advance by the many urban planners, engineers and computer scientists working in the Masdar City project. In this case, the machine literally follows what humans have fixed beforehand.

Masdar City's PRT was developed in 2009. More than ten years later, in the 2020s, the Emirati city is, like many other cities from all around the world, experimenting with a new form of urban transport: autonomous cars. An autonomous car shares a common trait with a PRT car: it does not have a driver and this is why, in common speech, autonomous cars are also called *driverless cars* and *self-driving cars*. However, an autonomous car functions in a very different way, compared to a PRT car. The former, as the name suggests, operates autonomously, while the latter moves in an automated manner. At the highest level of autonomy (level 5), humans are completely out of the loop, with the car driving itself in a real-life urban environment. Who or, in this case, what is steering the wheel is a non-biological intelligence possessing the main capabilities of AI, examined in the previous section. First, a car animated by an AI, has the ability to acquire data. It can sense the surrounding environment through a combination of different sensors, such as a video camera (detecting other vehicles, pedestrians and traffic lights), a radar (also monitoring the position of other vehicles) and a lidar system (sending laser pulses, 360-degree, which collide with surrounding objects and then reflect back, thus making the autonomous car understand what is around it). In addition, autonomous cars can share information among each other, and be fed with *big data* in the shape of, for example, detailed city maps and accurate weather forecasts.

Second, referring back to the common traits of AIs explained above, an AI-driven car would have the capacity to make sense of the acquired data. Sensing the colour red in traffic lights, for instance, would be translated into the concept of *stop*, while a green light would be interpreted as *go ahead*. Third, an advanced autonomous car would be capable of handling uncertainty and unexpected challenges. This is an essential skill to possess in cities which, as complex systems *par excellence*, continuously present out-of-the-ordinary situations like the one that occurred in Tempe (Arizona) on the 18 March 2018. A woman was crossing a road, walking outside the designated pedestrian crosswalk. Furthermore, she was pushing a bicycle laden with shopping bags. A prototype Uber autonomous car which was being tested on the same road, was not able to deal with that unusual situation. It hit the woman, killing her: the first pedestrian fatality caused by an autonomous car (see also Stilgoe, 2019). This accident evidences not only the complexity of the challenges that autonomous cars equally pose and face. It also puts emphasis on where they operate. Cars controlled by an artificial intelligence, like most urban artificial intelligences, act in real-life environments. This is a crucial aspect of urban AIs, as well as a defining feature of autonomy in the management of cities. *Real-life* means that there are actual human lives involved. The context in which the technology is employed is not a laboratory or a confined research facility, but a real city which, as such, is populated by humans. Drawing again on the example of Masdar City, the PRT car, as an automated technology, operates in an undercroft along tracks where the presence of people is not allowed. The PRT has its own environment, designed and built exclusively for the machine, so that it can endlessly repeat its pre-programmed tasks without interferences. On the contrary, autonomous cars share the same environment as human beings, thereby becoming an integrant part of the city. Within this lively environment, they have to make important decisions about what is right and what is wrong, and then act accordingly in an unsupervised manner.

The sharing of urban spaces between biological and non-biological intelligences has already taken place on many streets. Autonomous cars are being integrated into the transport portfolio of numerous cities, and it has been estimated that they will become the dominant form of urban transport by 2040 (Bansal and Kockelman, 2017; Milakis et al., 2017; Talebian and Mishra, 2018). Their employment is common in smart-city initiatives, as part of experiments in alternative urban mobilities, but it is rapidly exiting the niche of urban experimentation, thereby entering the mainstream. This escalation follows one of the typical dynamics of urban experiments observed in Chapter 1. Experiments in autonomous transport take little steps in small-scale testbed sites and then expand, embracing whole cities and states. Several countries, such as the US, UK, Netherlands and Singapore, for example, have released policies at the national scale to accelerate the deployment of fully autonomous vehicles on every road in every city (Government of the Netherlands, 2018; Government of the United Kingdom, 2018; Singapore Smart Nation, 2018; US Department of Transportation, 2017). This political push for an autonomous urban transport is shaking the city, in synergy with advertising campaigns carried out by the private sector (car manufactures and ridesharing companies, in particular), which promote AI as the future of urban mobility. From a historical perspective, the double political and economic force that is spreading the diffusion of autonomous cars in the city, resembles the advent of the early cars. In the 1920s, especially in North America and Europe, governments embraced the idea that the automobile was vital for economic growth and, together with powerful car manufactures like Ford and General Motors, they made the car the main form of urban transport in just a couple of decades (Hall, 2002; Sheller and Urry, 2000).

From a social point of view, the city is not a passive receiver of urban artificial intelligences and, in the case of autonomous cars, people are reacting differently according to different interrelated sociological, psychological, economic and cultural variables (Acheampong and Cugurullo, 2019). The age, the income and the education of the individual, the extent to which a person is in favour of technological innovation, the influence of significant others (family, friends and work colleagues) and personal beliefs in the sustainability of autonomous transport are examples of these variables (Acheampong and Cugurullo, ibid.; Daziano et al., 2017; Kyriakidis et al., 2015; Lee and Mirman, 2018). On the one hand, safety concerns are strong particularly among vulnerable road users, such as pedestrians and cyclists, who tend to be afraid of collisions (Kyriakidis et al., 2015; Penmetsa et al., 2019; Taeihagh and Lim, 2019). More in general, many are scared of what is a radically different urban transport technology: a car controlled by a non-biological intelligence. Studies show that numerous people fear that the AI might malfunction and are sceptical about its presumed ability to sense and handle the complexity of urban environments (Cugurullo et al., 2020; Hulse et al., 2018; Pettigrew et al., 2019). Others simply dislike AI and do not want to see it included in their mobility (Wang et al., 2020). Such scepticism and aversion go beyond the autonomous car, and are symptomatic of a broader resistance which some individuals manifest against the growing control of AIs over cities. However, on the other hand, these concerns and anxieties do not necessarily diminish people's interest in autonomous urban transport and, above all, the intention to use autonomous cars. Recent surveys on public attitudes towards AI-driven cars clearly indicate that while AIs are feared, most of the respondents are nonetheless willing to employ an autonomous car as their primary means of transport (Cugurullo et al., 2020; Hulse et al., 2018).

The autonomous car is not the only type of AI which is making its way into the city. Another category of urban artificial intelligence is represented by *robots* which have become a recurring presence in urban spaces. A robot is an artefact that, combined with AI, can sense the surrounding environment and interact with it in an autonomous manner (Russell and Norvig, 2016). Autonomous cars and robots are different urban artificial intelligences because the former category operates in one specific domain (transport) and, despite heterogeneous models, brands and sizes, the design remains fundamentally the same (an empty interior to accommodate passengers and wheels to move on urban roads). Robots instead exist in many shapes and forms, ranging from unmanned air vehicles (commonly called *drones*) to humanoid machines resembling the human body, and from almost invisible *nanobots* to *androids* barely distinguishable from a real person (Russell and Norvig, ibid.). These diverse types of robots are employed across a variety of urban domains, including customer service, retail, hospitality, education and security, as well as in the operation and maintenance of the infrastructure of the city (Macrorie et al., 2020; Tiddi et al., 2019; While et al., 2020). Like autonomous cars, advanced robots act in a real-life environment. Service robots, for example, often operate in the frontline, interacting and communicating directly with customers (Mende et al., 2019; Wirtz et al., 2018). The AI is not confined to a secluded space, but rather integrated into places, such as restaurants, where human and artificial intelligences are constantly engaging with each other (see Figure 7.2).

Artificially intelligent robots become part of the city by replacing a fundamental element of the city itself: humans. As explained in Chapter 1, from an Aristotelian perspective, the city has a social dimension made of all the people who inhabit it. Cities are composed of diverse individuals sharing the same built environment where they assume a variety of roles, acting in many different ways. AI, in the shape of robots, is entering cities by taking up roles and accomplishing services which have traditionally belonged to the human being. This process can be partly

FIGURE 7.2 A service robot in action at a restaurant in Rapallo (Italy). The gaze of the customers is directed elsewhere. In this place, the robot has become the new normal and, as such, unworthy of humans' attention.
Source: Stefano Mazzola

explained by focusing on the meaning of the word *robot*, which sheds light on the nature of robots. The term *robot* was introduced in 1920 by Karel Čapek, a Czech sci-fi writer. In his play *R.U.R.* (Rossum's Universal Robots), a robot is an artificially created person, so called after the Czech word *robota* meaning *forced labour* (Čapek, 2011). Robots are thus fundamentally workers. On the one hand, their technological components are today extremely diverse. In terms of mobility, for instance, robots can fly by means of wings and propellers, walk on legs or move using wheels. On the other hand, however, their core function is surprisingly homogeneous: to perform labour. This quality makes them highly compatible with the city which is a pivotal environment in the distribution and fulfilment of labour. It also makes them similar to humans since many robotic activities are originally human activities.

Together, autonomous cars and robots represent the most visible and relatable face of AI in the city. They are visible in the sense that they have an evident materiality which is observable when present in the built environment. Moreover, these urban artificial intelligences are relatable, due to the familiarity of their shapes and roles. An autonomous car is an automobile operated by an artificial intelligence. The driver (a non-biological intelligence) is considerably unusual, but the design of the car as well as what the car does have not changed substantially. Similarly, many types of service robots have an anthropomorphic appearance, in line with the kind of body structure that all human societies would be familiar with: they have two arms and one head, their posture is erect and their voice possesses a human-like tone. Service robots operating as waiters, chefs or policemen, for example, also carry out human activities. This does not mean that robots are universally accepted and no tension exists between them and humans. On the contrary, these similarities and affinities can cause problems because the diffusion of humanoid robots takes place in the city, mainly at the expense of the human being whose role is taken up by an intelligent machine. However, the same similarities and affinities are precisely what is facilitating the penetration of robotic technologies in the city. Many city-dwellers simply do not resist what they are already familiar with, and are not hostile to the idea of having their food prepared and served by a machine looking and functioning approximately like a human.

Beyond the above examples, urban AIs can assume unfamiliar and elusive shapes. This is the case of *city brains*, the third category of urban artificial intelligence. Alibaba's City Brain, introduced in Chapter 5, is emblematic. City Brain is a programme originally created for autonomous traffic management (Alibaba, 2020; Curran and Smart, 2020). It is an artificial intelligence capable of sensing the surrounding environment and making decisions across multiple domains in an unsupervised manner. More specifically, the environment that City Brain senses is the entire city. In this respect, the scale is much greater than that of an autonomous car or a service robot. The realm of City Brain is not a street or a restaurant, but whole urban settlements like Hangzhou, Suzhou, Chongqing, Macao, Guangzhou, Wuzhen, Shanghai (China) and Kuala Lumpur (Malaysia) where it is currently operational. A city brain acquires data primarily by means of hundreds of cameras distributed across the city. In addition, it can learn about the city by processing large data sets or, in other words, *big data* that Alibaba's computer scientists install and make available to it. Its process of learning is therefore both active and passive. A city brain, as Alibaba's example shows, learns by seeing and by being fed with big data. Its main operational domain is traffic, but Alibaba is expanding the agency of City Brain to other key urban realms such as safety, health, urban planning and governance. Within the domain of traffic, City Brain controls traffic lights, thereby controlling the flow of cars and people in the city and, ultimately, controlling urban mobility. Its decisions are unsupervised not in the sense that nobody is monitoring them. Computer scientists working for Alibaba are present, but they cannot dictate what the AI will do. A city brain functions in a hyper complex and ever-changing system: the city. In this context, it acts autonomously without following any pre-defined course of action, because it cannot know in advance exactly what is going to happen. Nobody does. As it was explained earlier in the third section, this is the key difference between automation and autonomy. The former takes place mostly in confined built environments where exceptions and variations are not allowed. The latter is applied in real-life scenarios where the unexpected is part of the everyday. Seen from this perspective, the realm of city brains is the unknown. Within this realm, a city brain has to make decisions in the presence of inevitable gaps in knowledge, facing persistent uncertainty.

The spatial dimension of City Brain is elusive. A city brain is, at the same time, everywhere and nowhere. It is ubiquitous because its range of action covers entire cities, but its artificial intelligence lies in an immaterial software which, unlike autonomous cars and service robots, is not confined to a specific and well-defined body. City Brain has some key physical components. Cameras, for instance, are the *sensors* by means of which the AI collects information, and their geographical distribution in the city and visual field represent part of the geography of the city brain. Traffic lights are the *actuators* that City Brain employs to act, with the aim of controlling urban traffic. Furthermore, the data collected by City Brain, as well as the effect of its actions, are processed through a *platform*. The images captured by City Brain's cameras and the real-time traffic patterns feed into a digital platform where information is analysed and visualized. For this reason, city brains can be understood as an instance of platform urbanism (see Barns, 2020; Caprotti and Liu, 2020; Leszczynski, 2020; Van der Graaf and Ballon, 2019). This is a faceless urbanism which is not immediately relatable to a single space, domain or actor and, yet, it has a single artificial intelligence whose autonomous decisions shape urban development, with considerable repercussions in terms of sustainability.

The (un)sustainability of the autonomous city

The proliferation of the urban artificial intelligences discussed in the previous section, evokes images of an emerging autonomous city. In the context presented so far in this book, the concept of *autonomous cities*, already employed by authors like Vasudevan (2017) and Norman (2018) to describe politically and economically independent cities, refers specifically to urban settlements managed and experienced by AIs which are capable of acting in an unsupervised manner (Cugurullo, 2020). The emergence of AI and autonomy in the operation of cities is a game-changer, inasmuch as it is creating urban spaces possessing non-biological intelligences which function while humans are out of the loop. Although, as noted earlier, autonomous technologies derive from automated technologies which have populated cities for many decades, the presence of artificially intelligent entities controlling part or the entirety of the built environment is an unprecedented phenomenon. The development of autonomous cities, therefore, poses pressing questions in terms of urban sustainability. AI is a *disruptive* technology, which means that it significantly alters the system where it operates (Batty, 2018; Greenfield, 2018; Kassens-Noor and Hintze, 2020; Yigitcanlar et al., 2020). How AI can transform cities and with what social, environmental, political and economic effects will be the focus of this section.

In the literature on AI, the impact that present and future artificial intelligences will have on human societies is widely discussed (Tegmark, 2017). This strand of the literature is notable for the fiery debate that sees many scholars either blessing or cursing the advent of AI, and offers an interesting starting point to reflect on the potential repercussions of autonomous cities. However, its overall utility is minimal when it comes to understanding the ramifications of the use of AI in cities because of how extreme the positions often are in the debate. At one end of the spectrum, foresight studies and predictions regarding the future of society in an age of AI take into account the *existential risk* hypothesis (Müller and Bostrom, 2016; Turchin and Denkenberger, 2018). This theory concerns the possibility of an advanced artificial intelligence with enough cognitive, political, economic or technological power to control the world, and whose motivations go against the interests of mankind. In this hypothetical context, such hyper powerful AI would either eliminate the human species, or establish a new world order which hinders human development. In essence, the existential risk hypothesis links progress in AI with an irreparable global catastrophe. From a philosophical point of view, this theory is actually sound. The orthogonality thesis, elucidated in the second section, posits that there is not necessarily a direct correlation between a certain level of artificial intelligence and the motivations that drive the actions of that artificial intelligence. Furthermore, as discussed above, an AI has several key features which are not present in human beings, and it is thus fair to assume that its interests might not be in sync with those of humans.

At the other end of the spectrum, AI is not seen as the terminal point of human development, but rather as the next step in human evolution. In order not to be overpowered by machines, humans merge with machines, thereby transcending the limits of biology. Kurzweil (2018) posits that human intelligence can be greatly enhanced by interfacing the brain with computers, which is a theory practised by Elon Musk via Neuralink (2020), an implantable brain-computer interface designed to augment the speed and scope of mental capabilities (Musk, 2019). The foreseen future is a society in which technologically enhanced humans have lost the control neither of the world, nor of AI: a superior *homo sapiens* has a vast array of artificial intelligences under control, and uses them to improve the conditions of mankind. For Lovelock (2019) such a scenario is unrealistic, because AI is the next step of life itself. His vision is that of a near future when super intelligent artificial entities quickly become the dominant species by virtue of their unparalleled capacity of processing information. These artificial creatures, unlike humans, are not organic. They are electronic and their materiality allows them to store and instantly share enormous amounts of data. Moreover, their bodies do not decay, and their digital consciousness can be eternally preserved in the shape of software. This is why, according to Lovelock (ibid.), the future of life on the planet and beyond is meant to be electronic. Following this narrative, humans (and organic life in general) become a closed chapter of Earth's history, but the end of the human race is optimistically peaceful, since Lovelock's AIs see the human being as a harmless neighbour with a common project: keeping the planet habitable.

The problem with these beliefs is that they are too Manichean. In other words, they present a dualistic image of the future in which AI is either benign or malign, and either in charge of the world or subjugated by humans. This black-and-white imaginary is misleading and, particularly in relation to urban sustainability, it fails to capture the broad range of issues that cities manifest every day, while waiting for a future planetary apocalypse which might never happen. Literature in geography shows that, although global phenomena harmful to human beings do exist, in practise their materialization tends to be uneven in time and space. Climate change is a prime example of these dynamics. There is substantial evidence of the transformations of the planet's weather systems (Bai et al., 2018; IPCC, 2014). Such global environmental changes, as Swyngedouw (2010, 2013) observes, have often been portrayed in the media as a universal upcoming Armageddon bringer of a species-wide destruction. However, empirical studies show a different condition, indicating that climate change is far from being temporally and spatially homogeneous. The evidence points to a variegated geography of climatic changes which are already damaging different environments, societies and cities (Ackerly et al., 2010; Jordan, 2019; Pecl et al., 2017; Romero-Lankao et al., 2018). There is no future global apocalypse, but rather a constant series of local and present repercussions.

The same logic applies to AI, cities and sustainability. When the focus is on the urban, thinking of artificial intelligence in general and absolute terms is erroneous. A single AI does not exist and is unlikely to exist in the next foreseeable future. Instead, there are and there will arguably be multiple and different AIs in multiple and different geographical contexts. AI technologies (like almost any technology) are the products of a constant process of modernization which is intrinsic to capitalist economies. As such, they change, evolve, multiply and spread globally in an uneven way, across a variety of scales ranging from the individual to the home, and from the building to potentially the entire city. The same urban space might host a plethora of diverse artificial intelligences, each one with its own social, environmental, political and economic impact. In addition, the application of the same AI in different urban spaces might have a myriad of heterogeneous outcomes. The emergence of AI in cities can be seen as a global phenomenon, but its practise and consequences should be understood as locally specific. Most importantly, from a sustainability perspective, the impact of AI on the city is gradual: it takes the shape of waves constantly hitting the shore, rather than the blast of a sudden Armageddon. This means that for cities to start feeling the influence of AI, an existential risk or the next grandiose phase of human evolution are not necessary at all. Artificial intelligence will have reshaped cities and urban living long before that. At the local scale, even a small AI can have significant repercussions, and waiting for a single omnipotent AI to take over the world or for humanity to control AI is futile. Such absolute visions lack nuance, and draw the attention away from the present and pressing urban changes that AIs are already triggering. Following this line of thought, the rest of the section seeks to reorientate the focal point of AI studies by focusing on existing urban AIs, their diffusion in the built environment and related sustainability implications. The enquiry begins with autonomous cars driven by an artificial intelligence.

The popularization of the autonomous car presents several possible scenarios; each one characterized by a different degree of sustainability (Acheampong et al., 2021; Sultana et al., 2019). An optimistic scenario in which autonomy and AI in urban transport lead to a sustainable urban development pictures the diffusion of autonomous cars together with the diffusion of ridesharing services à la Uber. Especially in large metropolitan areas, people are generally open to the idea of sharing an autonomous car, instead of owning one, and this attitude has the potential to decrease car ownership, thereby ultimately reducing the number of cars operating in cities (Fagnant and Kockelman, 2014; Firnkorn and Müller, 2015; Haboucha et al., 2017; Iacobucci et al., 2018). For example, it has been estimated that a single shared autonomous car can replace up to 11 conventional cars and four taxis (Alonso-Mora et al., 2017; Fagnant and Kockelman, 2018; Maciejewski and Bischoff, 2018). With more shared autonomous cars and, overall, fewer cars on the road, computer scientists working in the field of simulation have calculated that traffic will substantially drop (Guériau et al., 2020; Hörl et al., 2019; Levin et al., 2017). This scenario implies three crucial savings from an environmental sustainability perspective: energy, carbon emissions and space.

First, decreasing the number of cars in motion means decreasing the quantity of energy that is needed to make them move. Second, with fewer cars being used in cities, less carbon is emitted globally. The transport sector is responsible for more than 20 per cent of the CO_2 that human societies produce every year, with the automobile playing a cardinal role (Laakso, 2017; Tian et al., 2018; World Bank, 2020). Sharing autonomous cars can decrease the magnitude of transport and, in so doing, curb global carbon emissions, especially when what is powering the AI and the vehicle is a renewable energy. Third, given that the design of the city depends on local levels of traffic and on how many cars circulate in the built environment, the reduction of these factors opens up the possibility of less car-centric urban designs. Contemporary cities reserve up to 80 per cent of their total area for cars (Robertson, 2017). This urban space comes in the shape of roads, vehicle lanes and parking spaces which, together, form the realm of the automobile. Shared autonomous cars would not eliminate cars from the city, but they could diminish their number, thereby making part of the space currently designed for cars obsolete (Narayanan et al., 2020; Soteropoulos et al., 2019; Zhang and Guhathakurta, 2017). Even a small reduction of the percentage of the built environment now reserved for cars, would provide urban planners, policy-makers and urban designers with a large amount of urban space which could be redesigned and repurposed (Duarte and Ratti, 2018). The design of the city in an age of autonomous urban transport would not have to be defined by a car-centric vision and could instead find inspiration in visions of ecological urbanism akin to those presented in Chapter 2. A superfluous parking lot, for instance, might be turned into a garden designed for therapeutic purposes or for urban agriculture. An unnecessary vehicle lane could morph into a bike line or a pedestrian street. In essence, the diffusion of shared autonomous cars, could make cities greener, healthier and, above all, places for people, rather than spaces for cars.

However, the opposite scenario is also possible: a scenario in which the autonomous car further exacerbates the environmental problems presently caused by normal cars, ultimately condemning cities to a perpetual condition of unsustainability. A key issue is that the prospect of productive onboard activities promised by autonomous transport, could lead to more and longer commutes. Hawkins and Nurul Habib (2019: 69) observe that 'the decrease in travel disutility' commonly associated with self-driving cars can prompt 'people to travel more frequently and across greater distances.' Essentially, travelling in a car might become so productive and pleasant to the point of increasing the demand for cars, as well as the amount and length of car trips. For instance, individuals could work while an artificial intelligence drives them autonomously from home to their office. AI and autonomy also have the potential to make the car not simply an autonomous means of urban transport, but also a mobile space for leisure. This is the case of the Volvo 360c model whose horizontal design mimics the design of a bed, to encourage people to rest and sleep during long routes. A Volvo 360c is, by design, a car, an office, a bed or a living room depending on the individual needs of the customer: 'it is designed to help you relax or party on your journey, with all the creature comforts you need' (Volvo, 2020: no page). The emphasis is on *you*. This is not a product meant to be shared, but to be owned and individualistically consumed.

Such a line of autonomous transport would oppose the philosophy of sharing discussed above, thereby fostering car ownership. In this scenario, people would buy more cars and use them more often, with commuting being seen as a benefit due to the productive time unlocked by the AI behind the wheel. Environmentally, the negatives would be direct: more carbon emissions would be produced, and more energy and space would be needed to power autonomous cars and allow their transit. Socially, the problems would be subtler. Car manufactures like Volvo assure that their autonomous cars will create free time, thus empowering the individual with a new temporal slot (offered by the AI driver) which can be spent in compliance with one's wishes. Yet, these are empty promises, since the developers of autonomous vehicles cannot guarantee that the time made available by the AI will actually be free time. Passengers on an autonomous trip might decide to spend every moment working on tasks that are part of their daily job. This choice would turn free time into work time, and further blur the line separating what is work and what is not (McCarroll and Cugurullo, forthcoming). According to Crary (2014: 15), contemporary societies are rapidly moving to a 24/7 temporality: a condition of continuous work, 'amid the dissolving of most of the borders between private and professional time.' Cars driven by artificial intelligences, by providing extra space and time for work, could be the next step towards a 24/7 society.

The second category of urban artificial intelligences examined in this chapter, robots, presents a different situation. In this case, the social implications of the diffusion of robotic technologies in cities are explicit and at the forefront of public and academic debates. Service robots threaten the social sustainability of cities in particular, by challenging the employability of citizens and questioning their very role. As explained in the previous section, the meaning of *robot* is intrinsically connected to the concept of *labour*. Service robots are *de facto* an additional workforce which is penetrating and filling the social dimension of the city. Marx (2004: 517) had already noted how the machine was a 'mighty substitute for labour and workers' and nowadays, as Del Casino Jr (2016: 847) points out, robots are 'a new class of machines' which is disrupting labour systems. Scholars lament how humans could be easily replaced by robots, remarking that 'few employment fields are immune' (Bissell and Del Casino Jr, 2017: 436). In practise, the robotization of urban labour implies two direct consequences: unemployment and humans being forced to find alternative jobs in sectors which might not be in sync with their interests or vocation (Loi, 2015; McClure, 2018).

The gravity of the above phenomenon becomes evident through the lens of the philosophy of Aristotle, introduced in Chapter 1. For Aristotle (2000), the city is not a place for gods, but rather for people who find and follow a vocation, seeking to achieve their inner potential. Cities are places of human development, where individuals understand who they are and eventually take up a role which is in harmony with their identity. The process whereby humans comprehend and, above all, realize their potential is, for Aristotle (2000, 2004), *happiness* which the Greek philosopher calls *eudaimonia*. Service robots operating in cities then, by replacing an increasing quantity of human roles, have the potential to deprive humans of their eudaimonia, thus decreasing the social sustainability of the city. For a sustainable urbanism, eudaimonia is an essential resource to cultivate and protect, inasmuch as it is hard to imagine a sustainable type of urban living, which is not conducive to happiness.

Eudaimonia is not the only urban resource threatened by robots. There are other aspects of robotics to consider and relate to urban development. Korinek and Stiglitz (2017) have approached the study of the spreading of robots and AI technologies from an economic perspective, exposing the strong tension that exists between human development and the development of autonomous robots. They underline that, in terms of reproduction, humans are slower than robots, due to the fact that technological evolution is much faster than biological innovation (Korinek and Stiglitz, ibid.). Put simply, the way humans employ resources like food, energy and space to generate more humans, does not change and improve as quickly as robots raise their number and enhance their quality. This thesis is supported by the so-called Moore's Law. As Bostrom (2017) observes, Moore's Law was originally formulated to claim that the number of transistors on a microchip doubles approximately every couple of years, but it is today commonly used to refer to the exponential growth of computational power in ICT, including artificial intelligences. Building upon Moore's Law, Kurzweil (2018) highlights several important characteristics of the exponential development of AI technologies. First, it is not linear, meaning that the pace of change does not continue at the current rate: 'exponential growth is deceptive. It starts out almost imperceptibly and then explodes with unexpected fury' (Kurzweil, 2018: 8). Second, the power of AI increases rapidly. By *power* it is here intended the level of the abilities of a given AI, such as the capacity to perceive the surrounding environment, extract concepts and deal with uncertainty. Third, while the power of an AI keeps increasing, the cost to produce that AI decreases. For Kurzweil (2018), there is therefore an exponential growth in the *price–performance* relationship of AI, with the technology getting better and better, simultaneously getting cheaper and cheaper.

The combination of Korinek and Stiglitz's (2017) thesis with the expanded version of Moore's Law, points to the following conclusion: the diffusion of robots will grow exponentially at a much faster pace than the reproduction of humans. The costs of manufacturing service robots and drones will tend to fall, while the skills of these autonomous technologies will improve. Conversely, human development will follow a more linear pattern. Qualitatively, human abilities will grow and, quantitatively, so will the number of humans on the planet, but not as quickly as the growth of robots. As Korinek and Stiglitz (2017) are economists, their ultimate conclusion pertains to labour, and stresses the risk of robots eventually outnumbering and outsmarting humans. From an urban perspective, however, there are other risks to take into account. Historically, one of the key drivers of urbanization has been population growth, with existing cities expanding and new ones being built to accommodate a growing number of urban dwellers. In the autonomous city, if Moore's Law continues its course, the majority of urban dwellers are likely to be, in the long run, robots instead of humans. The exponential production of robots would then increase the production of the built environment, as more and more urban spaces and infrastructures would be needed to accommodate an ever-increasing robotic population. In such a scenario, the environmental externalities are severe. *Planetary urbanization*, discussed in Chapter 1, becomes less of a figure of speech to explain how the metabolism of cities affects the entire planet, and more literally the covering of most of the planet's surface with urban mass. The biomass of Earth (already declining) is increasingly replaced by artificial spaces. Moreover, unless the robotic population is sustained by renewable energies, its presence exponentially amplifies global carbon emissions. Even in an optimistic scenario in which robotic technologies are universally powered by clean energy, service robots and drones are ultimately artefacts whose production requires critical raw materials and whose disposal generates e-waste (Dauvergne, 2020). Their exponential growth cannot thus lead to urban sustainability.

Finally, with regards to city brains, the spectrum of their agency is such that important distinctions must be made to evaluate the sustainability impact of this large-scale urban AI. As mentioned earlier, Alibaba's City Brain acts in the realms of transport, safety, health, urban planning and governance, which, although interconnected, present different characteristics and challenges. The artificial intelligence developed by Alibaba has to deal with a broad range of situations, contrary to autonomous cars and service robots which engage with a specific task within a single context. In more practical terms, as the developers of City Brain state, the AI has the ability to 'predict what is going to happen next' (Zhang et al., 2019: 2). This is a skill whose meaning and implications can vary significantly according to the urban realm in which it is put into practise. In the case of urban transport, *what is going to happen next*, or simply *the future*, ranges from a couple of minutes to over 24 hours (Zhang et al., ibid.). Drawing upon meteorological data, a city brain, for example, can calculate today that tomorrow, due to adverse weather conditions, in a certain area of the city the possibility of car accidents will substantially increase. Based on this prediction, the AI could act on the traffic lights by increasing the yellow interval to give vehicles more time to stop at dangerous junctions.

The above example shows a straightforward application of a city brain which deals with uncertainty by collecting information about likely upcoming meteorological conditions and traffic patterns, and then acts to achieve the best possible outcome, where the meaning of *best* is in this case unambiguously understood as the reduction of traffic and car accidents. According to Alibaba's data analysts, in some of the cities where City Brain operates, travel time has decreased on average by 8 per cent and the intensity of traffic congestion has dropped by 15 per cent (Alibaba, 2019; Zhang et al., ibid.). These reductions correlate with reductions in carbon emissions and energy use, showing that, specifically in urban transport, the autonomous management of the city can foster environmental sustainability. However, in other urban realms, the actions of City Brain are exposed to substantial conceptual and technological challenges. There is a stark difference, for instance, between concepts such as *traffic congestion* and *road safety* and ideas like *justice* and *equity*. The first two concepts are easier to grasp. In the field of transport, as the previous example illustrates, a city brain can process their meaning and act accordingly. The city brain would identify congestion when perceiving long trip-times and vehicles queueing, and deem a road safe based on the number of car accidents that have occurred in that space. *Justice* and *equity* instead are more ambiguous ideas which are harder to define, quantify and measure. They also go beyond the domain of transport and underpin multiple aspects of the governance of cities. Chapter 5, for example, has evidenced how sometimes even a top-level policy-maker, in charge of a smart-city agenda, does not have a precise understanding of *equity*. In urban governance, artificial intelligences like City Brain engage with abstract ideas which are unclear to human intelligences in the first place, eventually acting in ethical grey areas.

The realm of urban planning (another sphere of influence of City Brain) encapsulates the broader challenge that urban artificial intelligences face: how the AI understands what is right and what is wrong. Urban planning is, in practise, a large-scale trolley problem in which unavoidable harm is constantly distributed in the city. Section two discussed how trolley problems undermine the hierarchy of the concepts at the foundation of AIs' decisions. An autonomous car operating according to the idea that it is right to protect human life, would struggle in a situation in which harming a human cannot be avoided. Similarly, an autonomous city controlled by a city brain with the same idea of *right* might struggle to plan a city that does not always hinder human life. This is because urban development tends to be a generally uneven phenomenon in which harms of different kinds are unequally distributed across the city. Contrary to the trolley problem, *harm* in urban planning does not imply the immediate death of somebody. Taking planning decisions does not suddenly unleash machines that run over people. Nonetheless, these are decisions which, in the long run, have a strong impact on human life. They would not be life or death sentences, like in the case of the trolley problem, but their consequences could profoundly affect the health, the well-being, the happiness and the longevity of individuals.

Planning for *housing* is a case in point. Chapters 5 and 6, for example, demonstrated how the planning of Hong Kong as a smart city is largely unequal. Only a minority of high-income workers can afford smart buildings, while a large share of the population is relegated to unhealthy accommodation. In this context, a city brain would have to make hyper complex ethical choices by taking into account that the development of a smart district would improve the lives of some, but hinder those of others. In such a grey area, it is unclear how the AI would ultimately act. Theoretically, its decisions would be based upon pre-designed performance measures which, if the AI in question is a *seed AI*, depend upon how the artificial intelligence is educated by its creators. In this instance, the seed AI would most likely follow the philosophy of urban development of whoever controls it, thereby potentially replicating human mistakes or adhering to successful practises of sustainable urbanism. In both cases, the AI's actions would be a reflection of what humans had previously done in the field of urban planning. Yet, the AI might also develop its own ideas of *right, wrong, good* and *bad*, and then translate them into visions of the *good city* and the *bad city*. Because of the orthogonality thesis, there is no guarantee that the vision of the good city cultivated and pursed by an artificially intelligent entity, will be in line with the needs and desires of humans.

Conclusions

Being that AI is a disruptive technology, its diffusion in the city is profoundly altering the city itself. This chapter has explored the emergence of a new kind of city resulting from the proliferation and integration of AI technologies into the built environment: the autonomous city. Autonomous cities are cities whose development and management is influenced by artificial intelligences operating in an unsupervised manner, with humans out of the loop. As the previous sections have shown, there are many different types of urban AIs characterized by different shapes, functions and levels of intelligence. They can coexist with each other or clash as opposing forces, just as their presence might occur together with older automated urban technologies, and their agency could intersect with that of human agents. In essence, the autonomous city is an extremely diverse category of the urban which, rather than causing a *tabula rasa*, is imposing a new urban layer on previously existing urban layers.

Such a stratum is made of urban artificial intelligences, like autonomous cars, robots and city brains, which are reshaping the social, economic, political and environmental dynamics of the city where they are integrated. Autonomous cities are then defined by AI, but AI is neither their sole component nor necessarily their main driver. Biological intelligences continue to play an important role, and so does the portion of the environment that does not possess autonomy. How a contemporary city metabolizes a new layer of urban AIs, varies on a case-by-case basis. Multiple forces are now clashing. Some urban stakeholders praise AI and try to accelerate its deployment. Others fear AI and attempt to resist it. This complex cultural, economic and political clash is unfolding globally, and the results are unlikely to be universal. Ultimately, a single dominant version of the autonomous city cannot exist as there are always geographical exceptions to take into account, ranging from places where large-scale AIs *à la* City Brain are banned to projects such as Neom and Beiyang AI Town: master-planned cities built *ex novo* to facilitate the integration of a vast array of AIs and managed by a ubiquitous artificial intelligence (Cugurullo, 2020; Neom, 2020).

The urban changes triggered by AI are also altering the sustainability of the city. This chapter has stressed that the magnitude of these changes can be considerable, and so their repercussions, but this does not mean that AI will lead to a catastrophic global event or to a new enlightened phase of human development. In relation to cities and sustainability, splitting the view of the future into extremes, as often happens in the literature on the societal impact of AI, is misleading. An existential risk, for instance, is certainly possible. However, the sustainability of cities is and will be substantially influenced by autonomous AIs, even if a malign AI does not take over. The alterations caused by artificial intelligences in the social, economic, political and environmental dimensions of cities are gradual and constant. The time to react to them is *now* and waiting for a hypothetical future Armageddon will only increase the unsustainability of the present. There is not one single place in particular where researching, questioning, discussing and critically engaging with urban artificial intelligences is needed. Every place and every context matter, because AI is spreading globally but unevenly and diverse urban spaces will filter it with different outcomes.

Overall, the rise of the autonomous city opens critical urban scenarios in which long-standing problems merge with pressing unprecedented challenges. On the one hand, the cases of Masdar City and Hong Kong suggest that urban artificial intelligences are emerging out of experimental eco and smart-city projects. On these terms, after the *eco-city* and the *smart city*, the *autonomous city* appears to be the latest trend in experimental urbanism, as well as the latest urban model supposed to achieve sustainability. As such, the autonomous city carries over the same issues that the book has previously discussed in relation to alleged eco-cities and smart cities. Although with the passage from *automation* to *autonomy* the technological apparatus of experimental cities is changing, an old string of urban problems remains. There is again a problematic emphasis on technology as a universal medium to enable a sustainable urbanism, a recurring presence in the making of cities of those companies that provide the technology in which faith is put (i.e. AI), a persistent one-dimensional understanding of sustainability which ignores social themes and includes environmental ones as long as they can be monetized, and an archaic vision of the city as a giant artefact ready to be optimized.

On the other hand, AI is a technology like no other which is pushing cities deep into the unknown. If smart-city technologies and initiatives have produced, for decades, data on the mechanics and metabolism of cities, then the passage to autonomy via AI suggests a further step: the ability to use data to extract concepts and make decisions without human supervision. In this sense, what differentiates the autonomous city from its predecessors is the capacity of thinking. This is a capacity which is still embryonal, but its impact and novelty should not be under-estimated. For the first time in history, cars and robots are capable of rudimentary thinking, and their skills and numbers are growing fast. With a city brain, an entire city can acquire the ability to think. It is thus fundamentally wrong and risky to treat the city as a passive artefact that can be optimized at will, while the city itself might soon have its own will. The risk lies in the fact that, although autonomous cities think and act, so far they lack the ethical norms of human intelligence, and this lacuna signs a type of urban evolution leading towards amoral cities.

When the development of AI intersects with the development of cities and human societies, questions of ethics are of paramount importance (Floridi et al., 2018; Taddeo and Floridi, 2018; Vinuesa et al., 2020). More specifically, the crux of the matter is how AI learns and understands what is right and what is wrong in ethically grey areas. This is a question cutting across the whole spectrum of urban artificial intelligences, from the autonomous car or the service robot in the position to have to distribute unavoidable harm among humans, to a city brain developing its own idea of the good city. Most importantly, this is a question which cannot be answered only by means of technological innovation. More advanced sensors and refined algorithms cannot discover and codify the meaning of good and evil. Simply because ideas, concepts and meanings are not discovered: they are discussed. Societies develop their ideals through philosophical enquiry and intellectual debate, and what ideals should drive the development of autonomous cities will be the central theme of the last chapter of the book.

References

Acheampong, R. A. and Cugurullo, F. (2019). Capturing the behavioural determinants behind the adoption of autonomous vehicles: Conceptual frameworks and measurement models to predict public transport, sharing and ownership trends of self-driving cars. *Transportation Research Part F: Traffic Psychology and Behaviour*, 62, 349–375.

Acheampong, R. A., Cugurullo, F., Gueriau, M., and Dusparic, I. (2021). Can autonomous vehicles enable sustainable mobility in future cities? Insights and policy challenges from user preferences over different urban transport options. *Cities*, 112, 103134.

Ackerly, D. D., Loarie, S. R., Cornwell, W. K., Weiss, S. B., Hamilton, H., Branciforte, R. and Kraft, N. J. B. (2010). The geography of climate change: implications for conservation biogeography. *Diversity and Distributions*, 16 (3), 476–487.

Alibaba (2019). City Brain now in 23 cities in Asia. [Online] Available: https://www.alibaba cloud.com/blog/city-brain-now-in-23-cities-in-asia_595479 [Accessed 10 November 2020].

Alibaba (2020). City Brain overview. [Online] Available: https://www.alibabacloud.com/et/city [Accessed 10 November 2020].

Allam, Z. and Dhunny, Z. A. (2019). On big data, artificial intelligence and smart cities. *Cities* 89, 80–91.

Alonso-Mora, J., Samaranayake, S., Wallar, A., Frazzoli, E. and Rus, D. (2017). On-demand high-capacity ride-sharing via dynamic trip-vehicle assignment. *Proceedings of the National Academy of Sciences*, 114 (3), 462–467.

Aristotle (2000). *The Politics*. Oxford University Press, Oxford.

Aristotle (2004). *The Nicomachean ethics*. Penguin Classics, London.

Aristotle (2009). *Prior analytics*. Oxford University Press, Oxford.

Armstrong, S. (2013). General purpose intelligence: arguing the orthogonality thesis. *Analysis and Metaphysics*, 12, 68–84.

Awad, E., Dsouza, S., Kim, R., Schulz, J., Henrich, J., Shariff, A., Bonnefon, J. and Rahwan, I. (2018). The moral machine experiment. *Nature*, 563 (7729), 59–64.

Awad, E., Dsouza, S., Shariff, A., Rahwan, I. and Bonnefon, J. F. (2020). Universals and variations in moral decisions made in 42 countries by 70,000 participants. *Proceedings of the National Academy of Sciences*, 117 (5), 2332–2337.

Bai, X., Dawson, R. J., Ürge-Vorsatz, D., Delgado, G. C., Barau, A. S., Dhakal, S., Dodman, D., Leonardsen, L., Masson-Delmotte, V., Roberts, D. C. and Schultz, S. (2018). Six research priorities for cities and climate change. *Nature*, 555 (7694), 23–25.

Batty, M. (2018). Artificial intelligence and smart cities. *Environment and Planning B*, 15, 3–6. doi:10.1177/2399808317751169.

Bonnefon, J. F., Shariff, A. and Rahwan, I. (2016). The social dilemma of autonomous vehicles. *Science*, 352 (6293), 1573–1576.

Bonnefon, J. F., Shariff, A. and Rahwan, I. (2019). The trolley, the bull bar, and why engineers should care about the ethics of autonomous cars [point of view]. *Proceedings of the IEEE*, 107 (3), 502–504.

Bansal, P. and Kockelman, K. M. (2017). Forecasting Americans' long-term adoption of connected and autonomous vehicle technologies. *Transportation Research Part A: Policy and Practice*, 95, 49–63.

Barns, S. (2020). *Platform urbanism: negotiating platform ecosystems in connected cities*. Springer, Berlin.

Bissell, D. and Del Casino Jr, V. J. (2017). Whither labor geography and the rise of the robots?. *Social & Cultural Geography*, 18 (3), 435–442.

Bostrom, N. (2012). The superintelligent will: Motivation and instrumental rationality in advanced artificial agents. *Minds and Machines*, 22 (2), 71–85.

Bostrom, N. (2017). *Superintelligence: Paths, dangers, strategies*. Oxford University Press, Oxford.

Čapek, K. (2011). *RUR & war with the newts*. Orion, London.

Caprotti, F. and Liu, D. (2020). Emerging platform urbanism in China: Reconfigurations of data, citizenship and materialities. *Technological Forecasting and Social Change*. 151:119690. doi:10.1016/j.techfore.2019.06.016.

Cave, S., Dihal, K. and Dillon, S. (Eds.) (2020). *AI narratives: A history of imaginative thinking about intelligent machines*. Oxford University Press, Oxford.

Clifton, J., Glasmeier, A. and Gray, M. (2020). When machines think for us: the consequences for work and place. *Cambridge Journal of Regions, Economy and Society*, 13, 3–23.

Cugurullo, F. (2020). Urban Artificial Intelligence: From Automation to Autonomy in the Smart City. *Frontiers in Sustainable Cities*, 2, 38.

Cugurullo, F., Acheampong, R. A., Gueriau, M. and Dusparic, I. (2020). The transition to autonomous cars, the redesign of cities and the future of urban sustainability. *Urban Geography*, 1–27.

Curran, D. and Smart, A. (2020). Data-driven governance, smart urbanism and risk-class inequalities: Security and social credit in China. *Urban Studies*. doi:10.1177/0042098020927855.

Dauvergne, P. (2020). Is artificial intelligence greening global supply chains? Exposing the political economy of environmental costs. *Review of International Political Economy*, 1–23.

Daziano, R. A., Sarrias, M. and Leard, B. (2017). Are consumers willing to pay to let cars drive for them? Analyzing response to autonomous vehicles. *Transportation Research Part C: Emerging Technologies*, 78, 150–164.

Del Casino Jr, V. J. (2016). Social geographies ii: robots. *Progress in Human Geography*, 40 (6), 846–855.

Descartes, R. (2003). *Meditations and other metaphysical writings*. Penguin, London.

Duarte, F. and Ratti, C. (2018). The impact of autonomous vehicles on cities: A review. *Journal of Urban Technology*, 25 (4), 3–18.

Fagnant, D. J. and Kockelman, K. M. (2014). The travel and environmental implications of shared autonomous vehicles, using agent-based model scenarios. *Transportation Research Part C: Emerging Technologies*, 40, 1–13.

Fagnant, D. J. and Kockelman, K. M. (2018). Dynamic ride-sharing and fleet sizing for a system of shared autonomous vehicles in Austin, Texas. *Transportation*, 45 (1), 143–158.

Firnkorn, J. and Müller, M. (2015). Free-floating electric carsharing-fleets in smart cities: The dawning of a post-private car era in urban environments?. *Environmental Science & Policy*, 45, 30–40.

Floridi, L., Cowls, J., Beltrametti, M., Chatila, R., Chazerand, P., Dignum, V., Luetge, C., Madelin, R., Pagallo, U., Rossi, F., Schafer, B., Valcke, P. and Vayena, E. (2018). AI4People – an ethical framework for a good AI society: Opportunities, risks, principles, and recommendations. *Minds and Machines*, 28 (4), 689–707.

Goodall, N. J. (2014). Ethical decision making during automated vehicle crashes. *Transportation Research Record*, 2424 (1), 58–65.

Government of the Netherlands (2018). Self-driving vehicles. [Online] Available: https://www.government.nl/topics/mobility-public-transport-and-road-safety/self-driving-vehicles [Accessed 10 November 2020].

Government of the United Kingdom (2018). Centre for connected and autonomous vehicles. [Online] Available: https://www.gov.uk/government/organisations/centre-for-connected-and-autonomous-vehicles [Accessed 10 November 2020].

Greenfield, A. (2018). *Radical technologies: The design of everyday life*. Verso, London.

Guériau, M., Cugurullo, F., Acheampong, R. and Dusparic, I. (2020). Shared autonomous mobility-on-demand: Learning-based approach and its performance in the presence of traffic congestion. *IEEE Intelligent Transportation Systems Magazine*, 12 (4), 208–218.

Haboucha, C. J., Ishaq, R. and Shiftan, Y. (2017). User preferences regarding autonomous vehicles. *Transportation Research Part C: Emerging Technologies*, 78, 37–49.

Hall, P. (2002). *Cities of tomorrow*. Blackwell Publishers, Oxford.

Hawkins, J. and Nurul Habib, K. (2019). Integrated models of land use and transportation for the autonomous vehicle revolution. *Transport Reviews*, 39 (1), 66–83.

He, J., Baxter, S. L., Xu, J., Xu, J., Zhou, X. and Zhang, K. (2019). The practical implementation of artificial intelligence technologies in medicine. *Nature Medicine*, 25 (1), 30–36.

Hörl, S., Balac, M. and Axhausen, K. W. (2019). Dynamic demand estimation for an AMoD system in Paris. *2019 IEEE Intelligent Vehicles Symposium*, IV, 260–266.

Horowitz, M. C. (2019). When speed kills: Lethal autonomous weapon systems, deterrence and stability. *Journal of Strategic Studies*, 42 (6), 764–788.

Hulse, L. M., Xie, H. and Galea, E. R. (2018). Perceptions of autonomous vehicles: Relationships with road users, risk, gender and age. *Safety Science*, 102, 1–13.

Iacobucci, R., McLellan, B. and Tezuka, T. (2018). Modeling shared autonomous electric vehicles: Potential for transport and power grid integration. *Energy*, 158, 148–163.

IPCC (2014). Climate change 2014: Synthesis report. Contribution of Working Groups I, II and III to the fifth assessment report of the intergovernmental panel on climate change. Core Writing Team, Pachauri, R. K. and Meyer, L. A. (Eds.). [Online] Available: http://www.ipcc.ch/pdf/assessment-report/ar5/syr/SYR_AR5_FINAL_full_wcover.pdf [Accessed 10 November 2020].

Jordan, J. C. (2019). Deconstructing resilience: why gender and power matter in responding to climate stress in Bangladesh. *Climate and Development*, 11 (2), 167–179.

Kanal, L.N. and Lemmer, J. F. (Eds.) (2014). *Uncertainty in artificial intelligence*. Elsevier, Amsterdam.

Kassens-Noor, E. and Hintze, A. (2020). Cities of the future? The potential impact of artificial intelligence. *AI*, 1 (2), 192–197.

Korinek, A. and Stiglitz, J. E. (2017). Artificial intelligence and its implications for income distribution and unemployment (No. w24174). National Bureau of Economic Research. [Online] Available: https://www8.gsb.columbia.edu/faculty/jstiglitz/sites/jstiglitz/files/AI_labor.pdf [Accessed 10 November 2020].

Kurzweil, R. (2018). *The singularity is near: When humans transcend biology*. Duckworth, London.

Kyriakidis, M., Happee, R. and de Winter, J. C. (2015). Public opinion on automated driving: Results of an international questionnaire among 5000 respondents. *Transportation research part F: traffic psychology and behaviour*, 32, 127–140.

Laakso, S. (2017). Giving up cars – The impact of a mobility experiment on carbon emissions and everyday routines. *Journal of Cleaner Production*, 169, 135–142.

Lee, Y. C. and Mirman, J. H. (2018). Parents' perspectives on using autonomous vehicles to enhance children's mobility. *Transportation Research Part C: Emerging Technologies*, 96, 415–431.

Leszczynski, A. (2020). Glitchy vignettes of platform urbanism. *Environment and Planning D: Society and Space*, 38 (2), 189–208.

Levesque, H. J. (2017). *Common Sense, the Turing Test, and the Quest for Real AI: Reflections on Natural and Artificial Intelligence*. MIT Press, Boston.

Levin, M. W., Kockelman, K. M., Boyles, S. D. and Li, T. (2017). A general framework for modeling shared autonomous vehicles with dynamic network-loading and dynamic ride-sharing application. *Computers, Environment and Urban Systems*, 64, 373–383.

Lin, P. (2016). Why ethics matters for autonomous cars. In Maurer M., Gerdes, J. C., Lenz, B. and Winner, H. (Eds.). *Autonomous driving: Technical, Legal and Social Aspects*, Springer, Berlin, pp. 69–85.

Loi, M. (2015). Technological unemployment and human disenhancement. *Ethics and Information Technology*, 17 (3), 201–210.

Lovelock, J. (2019). *Novacene: The coming age of hyperintelligence*. Allen Lane, London.

Maciejewski, M. and Bischoff, J. (2018). Congestion Effects Of Autonomous Taxi Fleets. *Transport*, 33, 971–980.

Macrorie, R., Marvin, S. and While, A. (2020). Robotics and automation in the city: a research agenda. *Urban Geography*, 10.1080/02723638.2019.1698868.

Marx, K. (2004). *Capital: Volume I*. Penguin Classics, London.

McCarroll, C. and Cugurullo, F. (forthcoming). *Social implications of autonomous vehicles: A focus on time*.

McClure, P. K. (2018). 'You're fired,' says the robot: The rise of automation in the workplace, technophobes, and fears of unemployment. *Social Science Computer Review*, 36 (2), 139–156.

Mende, M., Scott, M. L., van Doorn, J., Grewal, D. and Shanks, I. (2019). Service robots rising: How humanoid robots influence service experiences and elicit compensatory consumer responses. *Journal of Marketing Research*, 56 (4), 535–556.

Milakis, D., Van Arem, B. and Van Wee, B. (2017). Policy and society related implications of automated driving: A review of literature and directions for future research. *Journal of Intelligent Transportation Systems*, 21 (4), 324–348.

Müller, V. C. and Bostrom, N. (2016). Future progress in artificial intelligence: A survey of expert opinion. In Müller, V. C. (Ed.). *Fundamental issues of artificial intelligence*. Springer, Berlin.

Musk, E. (2019). An integrated brain-machine interface platform with thousands of channels. *Journal of Medical Internet Research*, 21 (10), e16194.

Narayanan, S., Chaniotakis, E. and Antoniou, C. (2020). Shared autonomous vehicle services: A comprehensive review. *Transportation Research Part C: Emerging Technologies*, 111, 255–293.

Neom (2020) Vision. [Online]. Available: https://www.neom.com [Accessed 10 November 2020].

Neuralink (2020). Neuralink. [Online] Available: https://www.neuralink.com [Accessed 10 November 2020].

Norman, B. (2018). Are autonomous cities our urban future? *Nature communications*, 9 (1), 1–3.

Nyholm, S. and Smids, J. (2016). The ethics of accident-algorithms for self-driving cars: An applied trolley problem?. *Ethical theory and moral practice*, 19 (5), 1275–1289.

O'Neil, C. (2016). *Weapons of math destruction: How big data increases inequality and threatens democracy*. Penguin, London.

Pearl, J. (2014). *Probabilistic reasoning in intelligent systems: Networks of plausible inference*. Elsevier, Amsterdam.

Pecl, G. T., Araújo, M. B., Bell, J. D., Blanchard, J., Bonebrake, T. C., Chen, I. C., Clark, T.D., Colwell, R.K., Danielsen, F., Evengård, B., Falconi, L., Ferrier, S., Frusher, S., Garcia, R.A., Griffis, R., Hobday, A.J., Janion-Scheepers, C., Jarzyna, M. A., Jennings, S., Lenoir, J., Linnetved, H. I., Martin, V. Y., Phillipa C., McCormack, P.C, Jan McDonald, J., Mitchell, N. J., Mustonen, T., Pandolfi, J. M. Pettorelli, N. Popova, N., Robinson, S. A., Scheffers, B. R., JustineD., Shaw. J. D., Sorte, C. J. B., Strugnell, J. M., Sunday, J. M., Tuanmu, M., Vergés, A., Villanueva, C., Wernberg, T., Wapstra, E. and Williams, S. E. (2017). Biodiversity redistribution under climate change: Impacts on ecosystems and human well-being. *Science*, 355 (6332), doi:10.1126/science.aai9214.

Penmetsa, P., Adanu, E. K., Wood, D., Wang, T. and Jones, S. L. (2019). Perceptions and expectations of autonomous vehicles – A snapshot of vulnerable road user opinion. *Technological Forecasting and Social Change*, 143, 9–13.

Pettigrew, S., Worrall, C., Talati, Z., Fritschi, L. and Norman, R. (2019). Dimensions of attitudes to autonomous vehicles. *Urban, Planning and Transport Research*, 7 (1), 19–33.

Richardson, R., Schultz, J. M. and Crawford, K. (2019). Dirty data, bad predictions: How civil rights violations impact police data, predictive policing systems, and justice. *New York University Law Review Online*, 94, 15–55.

Robertson, M. (2017). *Sustainability principles and practice*. Routledge, London.

Romero-Lankao, P., Bulkeley, H., Pelling, M., Burch, S., Gordon, D. J., Gupta, J., Johnson, C., Kurian, P., Lecavalier, E., Simon, D., Tozer, L., Ziervogel, G. and Munshi, D. (2018). Urban transformative potential in a changing climate. *Nature Climate Change*, 8 (9), 754.

Russell, S. J. and Norvig, P. (2016). *Artificial intelligence: A modern approach*. Pearson Education Limited, Harlow.

Sheller, M. and Urry, J. (2000). The city and the car. *International Journal of Urban And Regional Research*, 24 (4), 737–757.

Singapore Smart Nation (2018). Self-driving vehicles: future of mobility in Singapore. [Online] Available:https://www.smartnation.sg/initiatives/Mobility/self-driving-vehicles-sdvs–future-of-mobility-in-singapore [Accessed 10 November 2020].

Soteropoulos, A., Berger, M. and Ciari, F. (2019). Impacts of automated vehicles on travel behaviour and land use: An international review of modelling studies. *Transport reviews*, 39 (1), 29–49.

Stilgoe, J. (2019). *Who's Driving Innovation?: New Technologies and the Collaborative State*. Springer Nature, Berlin.

Stone P., Brooks R., Brynjolfsson, E., Calo R., Etzioni O., Hager G., Hirschberg J., Kalyanakrishnan S., Kamar E., Kraus S., Leyton-Brown K., Parkes D., Press W., Saxenian A., Shah J., Tambe M. and Teller A. (2016). Artificial intelligence and life in 2030. One hundred year study on artificial intelligence: Report of the 2015–2016 study panel, Stanford University. [Online] Available: http://ai100.stanford.edu/2016-report [Accessed 10 November 2020].

Sultana, S., Salon, D. and Kuby, M. (2019). Transportation sustainability in the urban context: A comprehensive review. *Urban Geography*, 40 (3), 279–308.

Swyngedouw, E. (2010). Apocalypse forever? *Theory, Culture & Society*, 27 (2–3),213–232.

Swyngedouw, E. (2013). Apocalypse now! Fear and doomsday pleasures. *Capitalism Nature Socialism*, 24 (1), 9–18.

Taddeo, M. and Floridi, L. (2018). How AI can be a force for good. *Science*, 361 (6404), 751–752.

Taeihagh, A. and Lim, H. S. M. (2019). Governing autonomous vehicles: emerging responses for safety, liability, privacy, cybersecurity, and industry risks. *Transport reviews*, 39 (1), 103–128.

Talebian, A. and Mishra, S. (2018). Predicting the adoption of connected autonomous vehicles: A new approach based on the theory of diffusion of innovations. *Transportation Research Part C: Emerging Technologies*, 95, 363–380.

Tegmark, M. (2017). *Life 3.0: Being human in the age of artificial intelligence*. Penguin, London.

Tian, X., Geng, Y., Zhong, S., Wilson, J., Gao, C., Chen, W., Yu, Z. and Hao, H. (2018). A bibliometric analysis on trends and characters of carbon emissions from transport sector. *Transportation Research Part D: Transport and Environment*, 59, 1–10.

Tiddi, I., Bastianelli, E., Daga, E., d'Aquin, M. and Motta, E. (2019). Robot–city interaction: Mapping the research landscape – a survey of the interactions between robots and modern cities. *International Journal of Social Robotics*, 1–26.

Turchin, A. and Denkenberger, D. (2018). Classification of global catastrophic risks connected with artificial intelligence. *AI & Society*, 1–17. doi:10.1007/s0014.

Turing, A. M. (1950). Computing machinery and intelligence. *Mind*, 59 (236): 433–460.

US Department of Transportation (2017). Preliminary statement of policy concerning automated vehicles. [Online] Available: https://www.nhtsa.gov/sites/nhtsa.dot.gov/files/documents/automated_vehicles_policy.pdf [Accessed 10 November 2020].

Van der Graaf, S. and Ballon, P. (2019). Navigating platform urbanism. *Technological Forecasting and Social Change*, 142, 364–372.

Vasudevan, A. (2017). *The autonomous city: A history of urban squatting*. Verso Books, London.

Vinuesa, R., Azizpour, H., Leite, I., Balaam, M., Dignum, V., Domisch, S., Felländer, A., Langhans, S.D., Tegmark, M. and Nerini, F. F. (2020). The role of artificial intelligence in achieving the Sustainable Development Goals. *Nature communications*, 11 (1), 1–10.

Volvo (2020). 360C. [Online] Available: https://www.volvocars.com/intl/cars/concepts/360c [Accessed 10 November 2020].

Wang, X., Wong, Y. D., Li, K. X. and Yuen, K. F. (2020). This is not me! Technology-identity concerns in consumers' acceptance of autonomous vehicle technology. *Transportation Research Part F: Traffic Psychology and Behaviour*, 74, 345–360.

While, A. H., Marvin, S. and Kovacic, M. (2020). Urban robotic experimentation: San Francisco, Tokyo and Dubai. *Urban Studies*, 0042098020917790.

Wirtz, J., Patterson, P. G., Kunz, W. H., Gruber, T., Lu, V. N., Paluch, S. and Martins, A. (2018). Brave new world: Service robots in the frontline. *Journal of Service Management*, 29, 907–931.

World Bank (2020). CO2 emissions by sector. [Online] Available: https://ourworldindata.org/co2-and-other-greenhouse-gas-emissions#co2-emissions-by-sector [Accessed 10 November 2020].

Yigitcanlar, T. and Cugurullo, F. (2020). The sustainability of artificial intelligence: An urbanistic viewpoint from the lens of smart and sustainable cities. *Sustainability*, 12 (20), 8548

Yigitcanlar, T., Desouza, K. C., Butler, L. and Roozkhosh, F. (2020). Contributions and risks of artificial intelligence (AI) in building smarter cities: Insights from a systematic review of the literature. *Energies*, 13, 1473. doi:10.3390/en13061473.

Yudkowsky, E. (2001). Creating friendly AI 1.0: The analysis and design of benevolent goal architectures. The Singularity Institute, San Francisco. [Online] Available: http://intelligence. org/files/CFAI.pdf [Accessed 10 November 2020].

Yudkowsky, E. (2008). Artificial intelligence as a positive and negative factor in global risk. In Bostrom, N. and Cirkovic, M. M (Eds). *Global catastrophic risks*. Oxford University Press, Oxford.

Zhang, J., Hua, X., Huang, J., Shen, X., Chen, J., Zhou, Q., Fu, Z. and Zhao, Y. (2019) City Brain: Practice of large-scale artificial intelligence in the real world. *IET Smart Cities*, 1, 28–37.

Zhang, W. and Guhathakurta, S. (2017). Parking spaces in the age of shared autonomous vehicles: How much parking will we need and where? *Transportation Research Record*, 2651 (1), 80–91.

8

EPILOGUE

The eclipse of urban reason

Introduction

Cities are human creations. They have always been made by humans to fulfil human needs and realize visions generated by human minds. On these terms, the reason why the city exists can be found in humanity, since cities are the product of humans' dreams, labour, qualities and instruments. This of course does not mean that non-human forces have not influenced urban development and that humans are perfectly in control of their urban creations. Cities have often evolved in unexpected ways, distorting the dreams of their creators. Nonetheless, without those human dreams there would be no city. The existence and shape of the city depend primarily on the genius, folly and hubris of mankind. However, this condition might change due to the proliferation of artificial intelligence (AI) in urban spaces and the formation of autonomous cities run by non-human intelligences. AI is a disruptive technology and so are its urban incarnations (Batty, 2018; Cugurullo, 2020; Golubchikov and Thornbush, 2020; Kassens-Noor and Hintze, 2020). Autonomous cars, robots and city brains have the capacity to transform the environment where they operate. Cars controlled by an AI do not need a driver, because the AI itself is the driver. A single AI can drive multiple cars, operating them together as part of a *platoon* which traverses the city as one mind in many bodies. Service robots can alter labour systems, assuming roles so far held by humans, and performing jobs that are nowadays unimaginable. City brains can govern a city by knowing everything that is happening inside it, and they can plan its development according to ideas cultivated by a non-biological intelligence. These transformations can be radical, as AI is a radical technology (Greenfield, 2018; Holton and Boyd, 2019). Urban artificial intelligences can colonize every aspect of cities and societies to the point of redefining them, thereby producing spatial formations and social organizations that are very different from what is now called a *city*. It is therefore crucial to proactively respond to the urban changes triggered by AI, instead of being shocked by them, in order to steer the development of autonomous cities towards urban sustainability.

The aim of this chapter is to critically examine how cities and societies are responding to the emergence of AI. The focus is on the main solutions that are being envisioned and implemented to face the challenges posed by AI. The narrative critiques the way cities are currently adapting to AI and sheds light on what ideas and actions are needed to align, in the near future, urban artificial intelligences with urban sustainability. In so doing, the chapter seeks to look forward into possible urban futures, while staying grounded in the present and looking back at the past. The future as well as the ever-changing present are, by nature, uncertain and what follows is not intended to be a final statement on future cities, but rather a preliminary and tentative investigation of potential AI-related urban futures that may or may not take place depending on a number of factors. The key factor that is under the spotlight is *ideological*. The book scrutinizes the ideas underpinning urban development in the age of AI, and theorizes what actual urban spaces and societies might grow out of these ideas. This is an exercise in critical theory meant to identify and analyze those ideas that are unsustainable, explain the mechanics through which they are emerging and causing urban problems, and indicate not a solution, but a path for reaching one.

This exercise will be carried out in connection with the material discussed in the previous chapters, which remerges here as lessons from the past and the present, containing signs of the future. From a sustainability perspective, the birth of AI in cities was not preceded by good omens. The passage from *automation* to *autonomy*, described in Chapter 7, has been driven by political economies often insensitive to matters of social and environmental sustainability. The second part of the book empirically explored the cases of Masdar City and Hong Kong. In the first case, the development of AI in the shape of autonomous cars, originates in the ambition of the developers and their business partners to find novel technologies to monetize. This ambition, in turn, originates in the objective of an authoritarian state, Abu Dhabi, to preserve its power and in that of tech multinationals to remain competitive in the global economy. In the second case, the genesis of AI technologies derives from the genesis of AI as a novel sector of the economy of Hong Kong. In this Special Administrative Region (SAR) of China, the local government sees great economic potential in the AI sector and has invested considerable capital to nurture it. In both cases, the benefits of urban experimentation have been unevenly distributed, ecological concerns have been scarcely taken into account, and the design of the built environment has lacked a consistent and holistic master plan. As a result, the urban experiments examined in this book have grown into a Frankenstein's monster made of disconnected and socially and environmentally dysfunctional elements.

In addition, both case studies display extended market analyses and shallow intellectual and philosophical efforts. The Masdar City project and Hong Kong's smart-city agenda are informed by extensive studies on their expected return on investment, while developers and stakeholders have made little or no attempt to engage with the meaning of *eco, smart* and *sustainable*. Emblematic is the formulation of Masdar City, allegedly an *eco-city*, whose core ideas differ considerably from the theories of ecological urbanism reviewed in Chapter 2. There is a substantial discrepancy between the business-centric Masdarian formula for urban development and the image of an ecologically healthy city inspired by the philosophies of social ecology and eco-socialism. Similarly, Hong Kong's policy-makers have been focusing on aligning smart urbanism with the production of in-demand technology, worried about attracting international companies to the SAR's business districts, but they have shown scarce interest in understanding if *smart* is actually *sustainable* and *for whom*. This issue is particularly severe in relation to the ethical challenges posed by AI. Chapter 7 stressed that urban artificial intelligences deal with complex questions of what is *right* and *wrong, ideal* and *undesirable* in real-life environments, which have a broad spectrum of impacts ranging from the safety of a citizen crossing a street where an autonomous car is in motion, to the long-term architectural development of a city planned by a city brain. These are questions which urge a collective philosophical debate, and the intellectual vacuum coupled with elitism, which is at the core of current eco and smart-city projects, is not promising. Hence the need to be critical of how urban experiments are paving the way for AI to enter the city.

Three sections form the conclusion of the book. First, the chapter delves into the ideologies that are emerging in response to the risks of AI. It discusses the theory of transhumanism as a model of development supposed to technologically enhance the mental capabilities of humans in a time when human intelligence risks being outclassed by artificial intelligence. Here the book explains the principles of transhumanism and builds upon them, developing the notion of *transurbanism* meant to signify the enhancement of urban spaces via AI. Both transhumanism and transurbanism are critiqued for unevenly distributing the benefits of superior technology and for undermining the very nature of what they should instead be preserving: the human being and the city. The argument advanced is that the way cities and societies are adapting to AI is fundamentally unsustainable. *Better technology* is how the risks of AI are mainly being addressed, but technology alone cannot solve deep-seated philosophical, planning and political problems. There is a lacuna in the way urban development is merging with development in AI in the sense that while technological innovation continues apace, philosophical inquiry, progress in planning and political engagement are lagging behind. At this juncture of the narrative, the book identifies a problematic divide in experimental urbanism, separating technological development from development in philosophy, urban planning and governance. Second, the chapter draws on the work of Max Horkheimer to inspect the roots of such a divide and examine its constant reproduction in capitalist societies. More specifically, it seeks to expand Horkheimer's study of reason, as originally presented in *Eclipse of Reason* (1947), by integrating key urban themes such as urbanization and urban sustainability into the philosophical system of the German scholar. The book develops a critical theory of experimental cities to illustrate the dynamics through which their development follows a track that prioritizes technological innovation to the advantage of the few, and disregards innovation in philosophy, planning and governance to the detriment of the many. Third, Horkheimer's lessons are employed to theorize how experimental urbanism can be improved, so that its technology becomes aligned with sustainable development. This is about critically questioning why urban experiments are developed in the first place and *to what end*. AI is a risky technology which is endangering cities, and its purpose must therefore be collectively discussed and evaluated by those who live in cities. The end of the book is about the end of the city. *End* as the final goal and destination that the city as a collective of human and non-human intelligences should be targeting. *End* as the decline of the city as a human space that, like Frankenstein's creature, might soon become an autonomous entity escaping human understanding and control.

Transhumanism and transurbanism

As Chapter 7 stressed, the development and diffusion of AI technologies in cities present a number of challenges. This section examines and critically evaluates two common strategies which are emerging in the twenty-first century, in response to the risks posed by urban artificial intelligences. Strategies that are connected to ideologies or, in other words, peculiar sets of ideas, beliefs, logics and ways of thinking (Eagleton, 2014). The first comes under the term *transhumanism*. The notion of transhumanism refers to a model of development seeking the biological evolution of the human being by means of scientific and technological innovation (Bostrom, 2005a). More specifically, transhumanism is about the enhancement of humans' mental and physical capabilities, and the overcoming of traditional human limits and burdens, such as the organic process of growing older, the perception of pain, and genetic limitations of the body and mind. This last aspect, in particular, is what the proponents of transhumanism usually highlight in debates over the future of human societies in the era of AI. Transhumanism's central argument is that the human brain's limited cognitive capacity can be greatly enhanced through the power of technology, in order to avoid a future scenario in which artificial intelligences outsmart and replace humans.

A transhumanist methodology can involve a variety of technologies, like nanotechnology, biotechnology and neurotechnology, depending on the human limit that is meant to be transcended. Methods and strategies include, for example, now common protein drugs boosting the anti-ageing capacity of the human body, thus challenging death itself (Lagassé et al., 2017). Particularly in relation to cognitive limits, the practise of transhumanism is exemplified by the invention of brain–computer interfaces supposed to improve the mental capabilities of humans, by adding the speed, storage capacity and computational power of a computer to the already existing strengths of the brain (see Neuralink, 2020). Such is the modification of the human being that, in transhumanism, humanity itself (intended as the quality of being human) is in question (Bostrom, 2005b; Ferrando, 2013). On the one hand, the transhumanist being gains new mental and physical capabilities. On the other hand, however, the transhumanist process of development forces the individual to lose qualities that are intrinsically human and have defined the meaning of being *human* since the birth of civilization. From a philosophical perspective, therefore, a poignant question is whether or not a transhumanist subject, such as a computer-interfaced person whose consciousness partly resides in a digital cloud that never ages, can be still considered as a human.

Besides this philosophically challenging issue, transhumanism also poses a more practical problem. Transhumanism in theory means that humans overcome their natural limits, thereby becoming a superior species, but transhumanism in practise clashes against a reality in which a homogeneous human species ready to be upgraded does not exist. As explained above, transhumanist methodologies are centred upon technology: this is a method of development which cannot be pursued unless biotechnologies, nanotechnologies and neurotechnologies are developed and implemented. Most importantly, transhumanism cannot be practised unless these technologies are bought and installed. Because transhumanist technologies have a cost and tend to be expensive, their diffusion in cities is uneven, with only a minority of wealthy individuals capable of purchasing them. In the long run then, the practise of transhumanism would lead to the creation of mentally and physically enhanced individuals, substantially superior to the people who do not have the economic capacity to afford transhumanist technology. In the case of brain–computer interfaces akin to Neuralink, for example, a rich elite would be able to purchase and install this novel neurotechnology. Such an elite, already economically superior, would become superior from a cognitive perspective too, in comparison with the non-mentally upgraded share of the population. In this example, the elite, now superior in both economic and cognitive terms, is not outsmarted by urban artificial intelligences. It remains competitive in the job market and keeps its role in the city. Conversely, those left behind economically and cognitively, find themselves competing against AIs and transhumanist elites, and their lives are at the mercy of superior forces.

The second strategy that is beginning to emerge around the world to reduce the risks posed by the autonomous city follows the same logic of transhumanism and applies it to urban development. The difference is that the subject is not the individual. It is the city. Given the strong conceptual similarities with transhumanism, this second response to the risks posed by urban AIs can be called *transurbanism*. If transhumanism is about the enhancement of the human being by means of superior technology, transurbanism concerns the enhancement of the city through superior urban technology and, more specifically, AI. This approach might seem initially counterintuitive, since the solution to the problem is identical to the problem itself: AI. In order to appreciate the lure of transurbanism, *teleology* offers a useful conceptual standpoint. Teleology is the belief that a given entity has an intrinsic potential and purpose, yet to be fulfilled (Aristotle, 1998). From a teleological point of view, the presence of problems is seen as a temporary condition likely to cease once the entity is fully developed (Aristotle, ibid). As noted in Chapter 7, some of the challenges presented by the autonomous city are due to the fact that AI technologies are not fully safe and reliable. AI is a technology under development and, as such, many of its features are perfectible. Chapter 7, for instance, discussed the case of Tempe's pedestrian fatality caused by an autonomous vehicle: a car driven by an imperfect AI kills a woman because it is not capable of sensing the pedestrian and stopping on time. This is an issue which can be read and then approached purely from a technological perspective. The urban AI in question cannot sense the surrounding real-life environment well and fast enough, and therefore its sensors (in this case, video cameras and the lidar system) must be enhanced. Transurbanism takes the same line of thought and adopts it in relation to every urban AI. This book has shown that artificial intelligence has multiple urban incarnations and resides in autonomous cars, robots and city brains. The book also explained that these diverse urban AIs share a series of key elements and skills. For example, they have the ability to acquire data (by sensing the environment or processing large data sets) and make sense of it (by extracting concepts from the acquired information). Moreover, they have the capacity to handle uncertainty and the ability to act in an unsupervised manner. In transurbanism, all these skills and abilities characterizing urban AIs are seen as *under development*. According to a transurbanist logic, a city filled with AIs is not a problem. Problems occur when urban artificial intelligences are imperfect. In essence, transurbanism finds the problem not in AI, but in the level of development of AI.

Like in transhumanism, immense faith is put in technological development. Transurbanism concerns the belief that urban AIs can be enhanced via superior technology. This view of urban development understands autonomous cars, robots and city brains as technologies that, compared to the state of the art, can *potentially* acquire larger volumes of data, better sense the surrounding environment, process information faster, extract less straightforward concepts and handle more complex uncertain situations, in a fully autonomous manner. *Potentially* in this context means that such advanced skills and capabilities are not present yet in contemporary urban artificial intelligences, but they will be present in the future provided that progress in disciplines like computer science and engineering continues. A fundamental aspect of transurbanism then is its focus on the potential of AI technologies: a focus which is sought through technological innovation as the medium to unlock the potential intrinsic to the autonomous city.

The similarities between transhumanism and transurbanism continue in the presence of both profound philosophical questions and critical practical problems. Transhumanism, by drastically modifying the human body and mind, alters the very quality of being human and, from a philosophical perspective, there is thus the questionable humanity of the transhumanist subject. Similarly, transurbanism radically alters the nature of cities and the essence of urbanism, by promoting the diffusion of enhanced autonomous artificial intelligences that experience and manage the built environment, ultimately overshadowing the role that the *human* plays in the *urban*. In philosophical terms, the question is whether or not autonomous cities transcend the boundaries of the traditional concept of the *city*, and if an urban development animated and shaped by an artificial intelligence still counts as *urbanism*. The infiltration of robots into society, for instance, is transforming a core dimension of the city. As Chapter 1 illustrated, cities have been traditionally characterized by a marked social dimension made of all the many individuals who experience the built environment engaging in a variety of social relationships. On these terms, a city is in part defined by the people who animate it. In the transurbanist city, however, the prospect is that of a plethora of enhanced robots populating the built environment. Some of these robots might be indistinguishable from humans and capable of performing most if not all social activities. As a result, several human activities taking place in the city or, in other words, *urban activities* could become robotic activities. Another possible scenario, considered in Chapter 7, sees the robotic population completely replacing large shares of the human population of cities. With its social dimension deeply reconfigured or, worse, trampled by another layer of robotic entities, a city undergoing a transurbanist process of development might lose an essential urban quality, thereby heading towards a spatial formation which cannot be called a *city*.

A similar conceptual conundrum concerns the enhancement of city brains. This is a type of urban AI that, as Chapter 7 has indicated, is already being used in the planning of cities. Such a phenomenon is unprecedented in the long history of the city. From an urban planning and design point of view, cities have essentially been human constructs. Technology has certainly aided human planners and designers in shaping cities, and this is particularly evident in smart-city projects where the management of urban services and infrastructures is often mediated by smart technologies (see Chapters 3 and 5). However, decisions about the shape and organization of the city have been traditionally made by human intelligences. A city brain questions this tradition, inasmuch as it consists of a non-biological intelligence making decisions which change the way cities are organized and shaped. As of this writing, an urban AI like Alibaba's City Brain functions together with human operators who are still involved in the planning process. Yet, given that transurbanism (just as transhumanism) is about continuous technological innovation, it is not hard to imagine a future city brain capable of autonomously planning the development of the city where it operates. In relation to the material discussed in Chapter 6, this would mean having an AI which develops its own urban equations. The AI in question would identify and define what ideas, concepts and themes should be included in the making of cities, and enforce their implementation. Consequently, the ideas at the foundation of cities planned by city brains might not necessarily be human ideas because their origin would not be a human intelligence. In this hypothetical but not improbable scenario, given the urban trends explored so far in the book, the very notion of *urbanism* is questionable since the form of the city and the relationship between its inhabitants and the built environment are shaped by a non-human intelligence.

From a practical perspective, the problems posed by transurbanism resemble those of transhumanism. In transhumanism, there is a severe issue of distribution, because biotechnologies, nanotechnologies and neurotechnologies are unlikely to be evenly developed, implemented, purchased and integrated, with the consequent unjust bifurcation of the human race split into enhanced and non-upgraded individuals. Similarly, in transurbanism, the distribution of advanced urban AIs will arguably be uneven and unjust. The reason is identical to the one presented earlier in the critique of transhumanism. Transhumanist and transurbanist technologies are dissimilar products in the sense that they do not have the same application, design and user. However, they are both technological products and, as such, they go through analogous processes of development, production, distribution and consumption. AI technologies are by default already expensive, and advanced urban artificial intelligences, capable of acquiring and processing data faster and better, will be even more expensive. Given that transurbanism sees urban artificial intelligences as *under development* and pushes for constant innovation in the field of AI, the practise of transurbanism means that there will always be on the market different urban AIs characterized by different levels of development and different prices. Particularly in the context of capitalist political economies, there will then be countries where only a minority of economically powerful cities will enjoy superior urban AIs, and cities within which solely an urban elite will have access to the best and most expensive artificial intelligences on the market.

The above theoretical scenario can be visualized in more pragmatic terms by referring to the practical examples of urban AIs exposed in Chapter 7. With regards to autonomous cars, for instance, transurbanism in theory promotes the development of a city whose transport sector is fully autonomous and accidents akin to the case of Tempe 2018 are impossible. The image portrayed by transurbanism is that of an autonomous city where cars driven by an artificial intelligence can perfectly sense the environment and react on time when out-of-the-ordinary situations arise. Fatalities are completely avoided because, should a situation similar to Tempe's events happen, the advanced lidar system of the autonomous car would spot the pedestrian long before the approach of the vehicle. The AI would then rapidly process the information, understanding the potential danger that lies ahead and recalculating its route so to avoid any collision. In relation to city brains, the image is that of a ubiquitous AI which controls urban traffic in an unsupervised manner. Congestion does not exist, because the city brain is processing data sets so vast that it can accurately predict when and where traffic jams might occur, and redirect in advance the flow of vehicles elsewhere. Car crashes due to adverse weather conditions do not exist either. The city brain has access to precise meteorological data produced in real-time, and it calibrates traffic lights to give vehicles more time to stop at dangerous junctions when the visibility is scarce or roads are slippery.

In reality, however, these ideal transurbanist images of development would crumble against a free-market economy in which different technologies are commercialized and unevenly distributed. The production of autonomous cars, for example, is not monopolized. There are many companies producing a variety of models of AI-driven cars characterized by heterogeneous designs, technologies and levels of autonomy. Consequently, there are and there will be (assuming no monopoly is formed) several types of autonomous cars whose characteristics, including the ability to handle out-of-the-ordinary situations, differ qualitatively. For the sake of argument, the assumption here is that there are only two types of autonomous cars called *alpha* and *beta*. This is of course an unrealistic hypothesis given that, in the current economic climate, capitalism enables fierce competition among numerous producers, but it serves the purpose of unveiling the crux of the matter. Alpha is a model characterized by superior AI technology, compared to beta which features less advanced software and hardware. More specifically, alpha has a better lidar system. It can spot pedestrians from far away, and its artificial intelligence processes information so quickly and vastly that even the most unlikely possibility of an accident is taken into account and avoided. This is not the case of beta. The typology of beta's technology is the same that alpha is sporting, but the degree of development is inferior. Beta too has a lidar system capable of spotting pedestrians, but its range is shorter than in alpha. Beta can also handle uncertainty, but not like alpha, and its weaker AI cannot predict what is going to happen next as well and as fast as alpha does. In essence, alpha is safer and more reliable and, due to its superior technology and capabilities, it is more expensive than beta. Not everybody can afford alpha and, therefore, autonomous cars are unevenly consumed and distributed as alphas and betas according to different levels of income. In urban terms, this means that there would be richer cities with a higher concentration of alphas, where the fatalities caused by autonomous cars are lower, compared to poorer cities where betas are more diffused. In addition, at the scale of the individual, the above situation implies that there would be people living in richer cities or districts where alphas are abundant, who would be safer when walking, in comparison with those living in places full of betas. Similarly, given that in the event of an accident passengers are at risk too, there would be individuals able to afford an alpha whose urban mobility would be safer than those who own a cheaper beta.

The same logic applies to other instances of urban artificial intelligence. A single model of city brain that is universally employed in every city is not a probable prospect. Like autonomous cars, city brains are technologies produced and developed by more than one company, which hit the market at different levels of development and prices. It is thus reasonable to picture the commercialization of alpha and beta city brains as the norm. Their characteristics and repercussions would mirror what was observed before in relation to autonomous cars, although on a larger scale. A city equipped with an alpha city brain would be less congested than a city whose traffic is controlled by a beta city brain. In this case, alphas have at their disposal bigger datasets as well as more cameras of a higher quality. Compared to betas, they can better predict the evolution of urban traffic, and avoid the concentration of vehicles on roads which are likely to become congested. In the long run, because of its superior traffic flow, alpha cities might gain considerable advantages in terms of sustainability. Economically, vehicles carrying goods and enabling services would reach their destination faster, thereby accelerating the city's economic activities. Moreover, thanks to the alpha city brain, automobiles are now in transit for less time. Consequently, they need less energy to complete their journey and emit less carbon. Therefore, environmentally, carbon emissions and energy consumption would decrease, together with a decreasing average travel time. Socially, the citizens of an alpha city, enjoying a superior mobility, could be more inclined to meet often and cultivate their social relations in person: an attitude that would collectively boost the social dimension of the city. Such advantages would be smaller or completely absent in beta cities.

These images of hypothetical future alpha and beta cities where diverse AIs, characterized by different levels of development, impact on urban sustainability in an uneven manner are grounded in the present. Alpha and beta cities, regardless of AI, already exist all around the world. So far, this section has taken existing technologies and pictured their upcoming deployment, according to contemporary models of urban development. The scenarios presented in this chapter are in line with the scenes previously illustrated in Chapters 3 and 4. The main difference between the present and the future is in the degree of technological development: not in the technology per se or in the way it is employed in cities. As the book has empirically shown, present eco and smart-city initiatives tend to produce fragmented spaces whose benefits and burdens are unevenly distributed. Masdar City can be seen as an alpha city accessible only to high-income workers. The Masdar City project is constantly pushing the boundaries of technological innovation, shifting from automation to autonomy, but most of its fruits are not shared with the rest of Abu Dhabi which, compared to the superior Masdarian built environment, is a beta city. Hong Kong's smart urbanism has generated a divided city in which smart technologies are available exclusively in premium locations. In the SAR, hyper expensive alpha smart homes and offices commercialized by the private sector and alpha smart districts built by the government to attract multinationals, are found next to non-upgraded beta buildings including rundown estates. Deep social fractures are visible in the urban models of the present and, in the near future, AI is set to exacerbate these long-standing issues.

The development of AI can become conducive to urban sustainability (in a manner that does not deepen already existing social divides) only if present urban models change first. This planning challenge, however, is connected to another practical challenge of transurbanism. Just like transhumanism, transurbanism too approaches problems from a technological point of view, seeking to overcome them by means of superior technology. Such a technocentric problem-solving attitude does not fully work when it comes to planning issues. Chapter 6 argued that current models of city-making are unsustainable because the ideas underpinning them are flawed. More specifically, eco and smart-city initiatives such as Masdar City and *Hong Kong Smart City Blueprint* fail to achieve sustainability, inasmuch as ideas like *justice, democracy* and *ecology* are not cultivated and then implemented by the developers who, instead, prioritize business-oriented themes. By advancing the concept of *urban equations*, the book exposed the need for a radical reformulation of contemporary planning strategies, with the central aim of questioning the very ideas that form the core of cities. This need and this aim cannot be fulfilled by technology alone. They require substantial intellectual and philosophical inquiry which is not what transurbanism is about.

A problem such as the circulation in cities of autonomous cars that are incapable of sensing pedestrians in out-of-the-ordinary circumstances is clearly different from the issue of urban equations centred around ideas like *profit, neoliberalism* and *elitism*. The former can be addressed through technological innovation in the shape of better lidar systems and cameras. The latter involves reflecting on the meaning of the good city and on what kind of urban future a given society aspires to. In the second case, better sensors and more computing power would not help much. Transurbanism's solution would be the enhancement of city brains, but this is a risky strategy. An advanced city brain in charge of planning a city and able to dis-cern what is *right* or *wrong* for that city is technically possible. This is a technology that, as mentioned in Chapter 7, is operative in the field of urban planning, meaning that the initial steps towards an autonomous urbanism have already been taken. Of course, from these initial steps getting to the end of the road might be a long way off, and the road itself is certainly full of obstacles, but none of them is in principle insurmountable. In fact, studies in computer science show that some AIs can currently play complex games involving making decisions about what they believe is the right or wrong strategy, in simulations comparable to decision-making in real-life urban planning (Crandall et al., 2018). Therefore, considering how rapidly AI technologies are evolving, the image of future city brains deciding upon what is *ideal* and what should be avoided is not unrealistic at all. The risk lies in the possibility that what a city brain deems *good* might not correspond to the good of humanity. This possibility was taken into account in Chapter 7 when the book unpacked the so-called *orthogonality thesis*, remarking that the interests of an artificial intelligence might diverge from human interests. It is thus crucial to always keep human intelligence in the loop in every urban project, with humans playing an active role in envisioning what they (and not an AI) believe is an ideal city.

However, the cultivation of collective human visions of development corre-sponds neither to the logic of transurbanism nor to the tenet of transhumanism. Both transurbanism and transhumanism have technological innovation as their common denominator, and pursue the endless development of superior technology without an overarching goal tailored to the whole human race. They also have three major flaws in common that prevent AI from making cities (transurbanism) and societies (transhumanism) sustainable. This section emphasized that the practise of transhumanism and transurbanism is highly individualistic in the sense that transhumanist and transurbanist technologies tend to be unevenly distributed, thereby benefiting only certain individuals and elites, rather than entire societies and cities. From an urban point of view, the first problem is a planning problem since it is caused by unjust strategies of urban development placing the benefits and risks of AI in an uneven manner. In addition, while transhumanist and transurbanist efforts of development focus on enhancing technology, little or no intellectual effort is put on questioning the purpose of technology. This is a critical issue in autonomous cities where autonomous cars, robots and city brains have to make important decisions based upon what they believe is *right* and *wrong*, according to values which might not be aligned with human values. The second problem is a philosophical problem that requires ethics or, in other words, the study of the principles that should drive human actions, human development as well as the development of human environments. Given that cities are the dominant human environment, ethics is crucial for defining the course of their development. Finally, to avoid the trap of individualism in which transhumanism and transurbanism persistently fall, ethics needs to be approached not by means of individual studies, but through a dialogue involving as many people as possible. Thus the third and last problem is essentially a problem of governance, because the challenge is to include every citizen in the formulation of a collective vision of the city and its future.

The above three problems are currently unresolved, and the lack of knowledge and action that they emanate is undermining the sustainability of cities in the age of AI. Urban artificial intelligences are disruptive technologies that are shaking the city's very essence, altering the way the built environment is managed, designed and experienced. Adapting to the drastic changes triggered by AI and facing the many challenges posed by autonomous transport systems, robots and city brains is vital. However, as argued in this section, the strategies that cities and societies are now following to counter the shockwaves propagated by AI will not lead to sustainability. Fundamental matters of urban planning, philosophy and governance are largely ignored in transhumanism and transurbanism and, unless they are incorporated into the development of autonomous cities, there is the possibility that the city of the future will cease to cultivate human flourishing. Future cities might stop being *human* spaces at all. To prevent an urban future that is not a human future and avoid the formation of autonomous cities where development in AI is not coupled with development in urban planning, philosophy and governance, the first step is to examine the origin of such lacuna. This is precisely the aim of the next section which draws upon the critical philosophy of the Frankfurt School in the attempt to build a critical theory of experimental urbanism. The book now turns to the theory of reason developed by Max Horkheimer (1895–1973), one of the fathers of the Frankfurt School, using it as a framework to understand why the creation of advanced artificial intelligences is not being followed by advanced philosophical inquiries, planning strategies and models of governance.

The eclipse of reason

Reason, for Horkheimer (2013), can be divided into two categories. The first type of reason is what the German philosopher calls *subjective reason*. This is the reason behind the actions of the individual: a force pushing a person to act and explaining why that person is acting in the first place. Subjective reason is a driver motivating the individual to live in a certain way, making specific choices and behaving according to them. Horkheimer (2013) points out that subjective reason, as the reason of the individual, is based on personal interests and gains.

The perspective taken into account is that of the self. In this sense, the individual would consider what is personally valuable, thereby prioritizing actions that would lead to any kind of personal advantage. The gain could be material (food or a house, for example) or immaterial (mental wellbeing or fame, for instance). The central aspect is that the gain in question is pursed by single individuals who directly benefit from it. As Horkheimer (2013) observes, there is nothing unnatural about this type of reason. On the contrary, for him such a way of thinking and acting is actually grounded in human nature, originating from people's primordial instinct of self-preservation. After all, the human being needs food and shelter to survive, and it is therefore normal for a person to consider and fulfil personal needs. Without subjective reason, humans would not take care of themselves and arguably the human race would not survive. In fact, Horkheimer (2013) himself remarks that there is also nothing wrong or problematic with subjective reason per se. However, the German scholar stresses that, since this form of reason focuses on the individual, it does not necessarily take into account what is outside and around the individual. More specifically, society, intended as a group of different individuals, falls outside the scope of subjective reason. Similarly, the environment, as the space where different people, societies and ecosystems are located, is not the same as the individual and thus could be ignored by subjective reason. Ultimately, the main risk highlighted by the philosophy of Horkheimer is that an extreme focus on subjective reason means an extreme focus on the needs of the individual, which, in turn, can translate into a complete disregard for everything that is out of the sphere of the individual, including society and the environment.

In the philosophical system developed by Horkheimer the second category of reason is called *objective reason*. Here the scale and scope of reason are much broader, compared to subjective reason. Horkheimer (2013: 2) describes objective reason as a 'force' which drives 'relations among human beings and social classes, social institutions and nature.' This is the reason that goes beyond the interests of the self by taking into account more and bigger entities. Objective reason is the reason of groups made of different individuals sharing the same space. It is the reason of societies and states. At the planetary scale, objective reason is the reason of humanity as a whole, and of the entire planet where humanity and other species reside. When objective reason is evoked, actions are not driven by personal needs, and the perspective under consideration is not that of the self. The needs are those of society, the environment, humanity and the planet, and the point of view cuts across many spaces and creatures, rather than being confined within the sphere of a single person. In essence, while subjective reason engages mostly with self-preservation and personal development, objective reason attempts to find an equilibrium between the interests of the self, and those of society and the environment.

For Horkheimer (2013), a model of development based on objective reason is necessarily inspired by so-called *higher concepts*. These are complex concepts seeking to capture needs, emotions and feelings that are not centred only on the self. In

Eclipse of Reason, Horkheimer presents *justice, happiness* and *equality* as examples of higher concepts. In so doing, he draws upon the philosophy of thinkers like Plato (2007) and Aristotle (1996, 2000) who tackled the question of development in relation to large socio-environmental systems considering but also transcending the self. On these terms, the ideas of Aristotle discussed in Chapter 1 are emblematic. Aristotle (2000) interprets the notion of development as a process composed of different stages, starting with single individuals who come together as a family. Different families then form a village, and eventually the union of several villages leads to the development of a city. This is also, in Aristotelian terms, the process of urban development which explains the origin and essence of cities. As noted in Chapter 1, in Aristotle's philosophy, happiness (which the Greek philosopher calls *eudaimonia*) is one of the key goals of urban development. In this sense, urban development is, from an Aristotelian perspective, a type of development inspired by a higher concept, because it seeks to produce happiness in the city.

Happiness exemplifies objective reason, inasmuch as it does not exclusively involve the self. For Aristotle (2000, 2004), cities should aim to generate happiness for every citizen. It is not the need of a single person that is taken into account. Instead, this concept captures the need of the city as a collective of individuals. Moreover, happiness transcends the self, because the individual alone cannot obtain happiness. In Aristotelian philosophy, happiness as eudaimonia signifies *human flourishing*. Humans develop as they discover and cultivate their inner potential. Aristotle (2000) argues that people cannot fully realize their potential by themselves. They need other people to help them, as much as they need urban spaces where key resources, services and infrastructures are made available. In a word, what humans need to flourish are *cities*. In addition to the traditional examples presented by Horkheimer, a contemporary incarnation of objective reason can be found in the concept of *sustainability*. Like justice, equality and happiness, sustainability is a higher concept since it concerns individuals not as standalone entities but as part of social groups which interact with the environment. A *sustainable development* is a type of development which pursues the needs of the individual only in connection with the needs of other individuals (including future generations) and with those of the non-human natural world. What counts is the anthropos as a species coexisting with other species in a variety of ecosystems.

Higher concepts, such as happiness and sustainability, are complex for a twofold reason. First, from a conceptual perspective they are difficult to understand and unpack. This issue becomes particularly evident when a higher concept associated with objective reason is compared with a non-higher concept related to subjective reason. Compared with the need for food, *justice*, for instance, is much harder to define. Establishing the meaning of *just* and *fair* presents bigger intellectual challenges than explaining the basic fact that the human body needs nutrients to produce energy, grow and repair. Furthermore, the energy needed by humans to breathe, move and activate essential human processes is universal. Conversely, what is just in one place could be seen as unfair in another location. Therefore, due to its context-dependent nature, *justice* requires more and deeper thinking across geographical spaces. Second, from a practical point of view, higher concepts are extremely hard to realize. *Happiness*, for example, unless it is pursued democratically in a way that leaves citizens free to be happy (or even unhappy) according to their desires and needs, risks becoming a form of violence against the subjective experience of the individual (Ahmed, 2010). *Sustainability* is another a case in point. Assuming that the enormous intellectual challenge to come up with a clear, specific and cross-boundary understanding of sustainability was overcome, there would be an even larger challenge of practising a global sustainable development. Considering the plethora of individuals, societies and ecosystems involved, this is a wicked problem that only a very sophisticated system of global governance can tackle, and certainly tougher than the problem of a single person looking for something to eat.

Horkheimer (2013) claims that there is an eternal tension between subjective reason and objective reason. These two types of reason are two different *sides* of human nature, meaning that potentially they can coexist. The reason driving the actions of a person, for instance, could be partly subjective and partly objective. This is a perfectly possible scenario because, to maintain the previous examples, somebody could seek food and shelter while still caring about issues of justice and sustainability. However, the German philosopher laments that in modern societies, subjective reason, the reason of the individual, largely overpowers objective reason, the reason of the whole. What he calls the crisis or eclipse of reason is a process unfolded in two steps. The first step consists of the absolute triumph of subjective reason and the consequent negation of objective reason. For Horkheimer (2013), this process ultimately dissolves the objective content of reason by eliminating the objective side of reason. A major implication of this phenomenon is that higher concepts, which are the domain of objective reason, stop being part of reason. Emptied of ideals like justice, equality and democracy, reason becomes for Horkheimer a void that can be filled with personal interests and exploited for personal gains.

This does not mean that higher concepts simply disappear from the landscape of human knowledge. The repercussions of the eclipse of reason are more complicated than that. Higher concepts continue to exist but mostly in a nominal way, thereby becoming empty words. *Sustainability* can be used again as an example to elucidate this condition. As several scholars point out, the concept of sustainability has often been emptied of any objective content and, through this process, it has become a black box open to many different understandings and interpretations (see Krueger and Gibbs, 2007). As an empty signifier meaning anything and nothing, *sustainability* has been frequently instrumentalized by politicians, policy-makers and private companies to pursue individual ends instead of collective interests (Brown, 2016; Swyngedouw, 2007). In the field of urban development, a number of studies show that the term 'sustainability' has been repeatedly exploited as a label to mask pro-economic growth agendas which have fostered social inequality and environmental degradation (Bouzarovski et al., 2018; Cugurullo, 2013; Imrie and Lees, 2014; McClintock, 2018; Raco, 2005; Whitehead, 2007). *Sustainability* can thus be present as a word which does not correspond to what is actually happening on the ground. Analogous reflections abound in classical and modern philosophy. Critical philosophers, in particular, have dug into the discrepancies between what exists nominally and what does not in practise. Chapter 1, for instance, introduced *correspondence theory* to critique statements that do not correspond to facts (Kirkham, 1992). Similarly, Foucault (2002) talks about *discourses*, stressing how language becomes a treacherous medium to establish and preserve uneven power relations. Regardless of the terminology adopted, the heart of the matter is that in the eclipse of reason higher concepts are found in the shape of words, statements, labels and discourses which are not necessarily employed to serve the purpose of the original higher concept.

The second step towards Horkheimer's eclipse of reason is 'the complete transformation of the world into a world of means, rather than of ends' (Horkheimer, 2013: 66). For the German philosopher, this process is due to the triumph of subjective reason and to its tendency to pursue primarily the development of means, intended as instruments which are designed and built to fulfil personal objectives. To better understand this aspect, it is important to remember the main traits of subjective and objective reason. The former focuses on the needs of the individual, while the latter engages with what large systems, like a society, a city, the environment or even the entire planet, need. Subjective reason targets self-preservation and self-gain, while objective reason follows higher concepts such as justice, happiness, democracy and sustainability. Horkheimer (2013) posits that, because of their diverse nature, these two types of reason tend to promote two types of development, which are outlined in *Eclipse of Reason* and expanded below. On the one hand, subjective reason, by seeking more and better instruments, fuels technological development and innovation. On the other hand, objective reason, by dealing with intellectually and logistically complex concepts, advances philosophical enquiries and studies in the fields of governance and planning. In the first case, the outcome is advanced technologies, devices and services. In the second case, the results are detailed visions of development, holistic planning strategies and sophisticated models of governance.

The above abstract theories reflect the previous practical examples. The basic personal need for shelter promotes the development of structures, infrastructures and technologies that provide cover and protection. Such impetus translates materially into, for instance, houses, energy grids and heating systems protecting the individual against cold weather. In addition, the same need drives the development of products and services which the individual can sell to generate the money that is necessary to buy a house, access the local energy grid and utilize a heating system. A similar line of thought applies to other personal needs like food, clothing and healthcare. The realization of a higher concept, however, is different. Achieving sustainability, for example, necessitates first and foremost defining what *sustainability* means. As discussed earlier, this is a very complex task as sustainability is a concept dealing with a multitude of individuals, societies and ecosystems, and comprehending several dimensions (social, environmental, economic, political and cultural spheres) intersecting with each other. Furthermore, as *sustainability* is related to the whole planet, the challenge is to develop an understanding that ideally speaks to every society, without losing detail and becoming a black box. In this regard, the Brundtland Report's *Our Common Future* and, more recently, the Sustainable Development Goals manage to capture the global scale of *sustainability*, but fail to deliver a detailed understanding of what exactly it means to be *sustainable* in a way that is unambiguous and unmanipulable. Addressing such a colossal intellectual challenge seriously implies pushing the boundaries of philosophical inquiry and social theory. Second, the achievement of *sustainability* is a target that requires development in planning and governance. At the scale of the city, an efficient and holistic planning system is an essential requisite for coordinating the many activities and projects supposed to protect the bioregion, decarbonize the built environment, create public spaces, provide sufficient housing and cultivate the eudaimonia of every citizen. At the scale of the planet, a system of global governance is a *conditio sine qua non* in order to mitigate and fix transnational problems, such as climate change, pandemics, water supply and the regulation of artificial intelligence.

As the two sides of reasons are not always mutually exclusive, the two types of development that they promote do not necessarily exclude each other. In relation to the examples just discussed, it is clear that the technological development underpinning the provision of shelter, can go hand in hand with the realization of a higher concept. Targeting *justice* in a city, for instance, is connected to the provision of housing which is not only a social and political issue. Housing has an evident technological aspect since it implies the material construction of homes and, as such, it can benefit from development in the fields of engineering, materials science and even computer science when smart-home technologies are taken into account. By the same token, the achievement of a higher concept like *sustainability* can be facilitated by technological innovation. The reduction of cities' carbon emissions aligns well with the development of generators of renewable energy, and AI can provide extra computational power to process extraordinary difficult global problems (Yigitcanlar and Cugurullo, 2020). Technology and philosophy are not natural-born enemies.

The problem is that, in the eclipse of reason described by Horkheimer, objective reason has annihilated subjective reason, and technological development thus reigns supreme without pursuing any higher concept. As the title of his book clearly indicates, the German philosopher depicts a total eclipse: the complete obscuring of the sun and the light it casts. It is an extreme event that Horkheimer uses as a metaphor to denounce a world in which he does not see justice anymore. A world dominated by reckless technological innovation, where philosophy struggles to illuminate a path towards just and democratic societies. In *Eclipse of Reason*, such fierce critique is directed against actually existing forms of development and social organization (a trademark of the Frankfurt School). In particular, Horkheimer (2013) critiques capitalism as a development model pursuing technological progress to improve the conditions not of the whole society or the environment, but those of an elite. In capitalist societies, *modernization* intended as the constant production of novel technologies and services, benefits directly the few individuals who control the means of production and gain capital by selling ever more and better products. Furthermore, modernization benefits the people who can afford its fruits and, put simply, have the economic capacity to pay for technological products and services.

This is the same line of criticism that has emerged in the empirical chapters of this book. Eco and smart-city projects have been critiqued for mainly being initiatives seeking the production of urban technologies that benefit an elite. In the Masdar City project and Hong Kong's smart-city agenda, higher concepts exist, but their presence is frequently only nominal in the shape of empty words, labels, statements and discourses. Masdar City is labelled as an *eco-city*, but what the developers are actually doing differs substantially from the theory of ecological urbanism discussed in Chapter 2, and is in sync with the laws of economics rather than with those of ecology. *Hong Kong Smart-City Blueprint* is, according to the local government, turning the whole SAR into a smart and sustainable region, whereas, in reality, smart technologies are being implemented only in a few selected premium spaces. The main difference between Horkheimer's critique and this book's case study research is that while the former presents a total eclipse, the latter illustrates an ongoing eclipse, suggesting that the process can be reverted. The cardinal point made by Horkheimer in *Eclipse of Reason* remains: technological development without any overarching vision of development is ultimately a blind form of development. Unless higher concepts cultivated by means of philosophical inquiry are taken into account, technological progress will be manipulated for personal gains, instead of serving the needs of society as a whole and the environment. As the previous empirical chapters have emphasized, so-called *eco-cities* and *smart cities* have been emerging as urban engines animating an endless production of novel technology which is now culminating, with the rise of AI, in the passage from automation to autonomy. This rapid process of technological innovation has impacted cities with little or no input from the objective side of reason. As a result, it has been exploited to support the business interests of elites and sustain unjust capitalist political economies. On these terms, the development of urban artificial intelligences is following a track already dictated by subjective reason, heading towards an urban future bereft of higher concepts. The solution to this issue, however, is not the elimination of smart and AI technologies. Technology per se is not the problem, exactly as subjective reason per se is not problematic. The problem lies in the absence of objective reason and related higher concepts. The solution, therefore, has to be the stimulation of objective reason and the integration of higher concepts into urban development. In this sense, adverting the eclipse of reason in cities means questioning why cities are built and developed in the first place. Specifically, in relation to experimental urbanism, what has to be urgently questioned is *to what end* AI is being integrated into the city. Together these questions call for a theory of urban reason which will be the focus of the next and final section of the book.

The ends

Understanding the reason why cities are built is a daunting task, rich in complexity. Similarly, clarifying why artificial intelligence is now being employed in urban development is unlikely to produce a straightforward explanation. Besides, there is more than a single *why* to consider. One, for instance, could explain the political, economic and social mechanics, the planning processes and the contextual pressures through which certain types of city emerge and then come to integrate AI. Unpacking this *why* is complex but not unfeasible. This has been a key goal of the book which has examined the genesis of eco and smart-city initiatives, discussing why these urban projects are developed and revealing how, in the passage from automation to autonomy, they begin to include artificial intelligences as part of their fabric and design. However, developing a clear understanding of the reason why these experimental cities *should* be built in the first place is a different and much harder task. It involves identifying a collective goal that gives meaning to the city in question and justifies the consumption of the extraordinary amount of resources necessary to sustain its metabolism. This second *why* goes beyond eco-cities, smart cities and autonomous cities and, at a higher abstract level, it can be extended to any spatial formation of urban origin. *Why cities?* is a structurally simple question, but its interrogation point leads to an enigma that very few scholars have attempted to explore in-depth. Clarifying why it is important and desirable to build cities and specifying what it is that cities are built for are questions carrying enormous weight.

These questions are currently unresolved but Horkheimer's theory of reason can shed light on them to discover possible answers. Whatever the reason is that experimental cities are developed for, it has to strike a balance between subjective and objective reason. Present eco and smart-city projects and their evolution into autonomous cities are influenced by the subjective side of reason and, as such, they tend to pursue almost exclusively the interests of individuals and elites. Higher concepts like *justice, democracy* and *sustainability* are completely absent or, at best, marginalized compared to business-related themes tailored around individualistic gains. Stimulating the objective side of reason is, therefore, essential to justify cities whose formation can be desirable to the whole population and the surrounding ecosystems. Higher concepts must be included in contemporary urban equations, and objective reason is a prerequisite for their cultivation and thus for the sustainability of cities.

While it is outside the scope of this study to thoroughly theorize why cities are built, the material discussed so far suggests the presence of an element necessary for a theory of urban reason, which future studies might use as a steppingstone. Cities need ends. In this context, the word *end* has a twofold meaning. First, *end* as an objective or a goal which underpins urban development. In this sense, urban reason is defined by the aim that should be achieved through the genesis and development of a city. For example, a city could be constructed in order to provide shelter for the entire population of a region, as it might grow to realize the eudaimonia of every citizen. In these cases, urban reason is the *even provision of shelter* and *widespread human flourishing* respectively. There could easily be more than one goal, inasmuch as certain aims spontaneously resonate with each other and exert a gravitational pull which attract other aims. The aim of providing *affordable housing*, for instance, is connected to the aim of achieving *justice*. Likewise, targeting *democracy* and *citizen participation* are goals whose successful accomplishment requires the promotion of *education*.

When these aims or ends are driven by higher concepts, they are most likely to have multiple geographical dimensions reflecting their broad scope. As explained earlier in this chapter, a higher concept is based on objective reason and goes beyond the narrow sphere of the self, thereby taking into account groups of individuals and different societies. Potentially, higher concepts can encapsulate the whole human race and even the entire planet with its innumerable ecosystems. A city seeking to achieve sustainability, therefore, could not concentrate exclusively on what happens within its territory. In relation to smart urbanism, for example, this book has put emphasis on some of the critical supply chains that cut across diverse geographical spaces, noticing how smart technology necessitates raw materials that are often extracted far away from where smart interventions occur. As a result, while one place is becoming smarter, at the other end of the supply chain, environmental degradation and labour exploitation spread. If *urban sustainability* is approached seriously as a higher concept, and not as an empty label or discourse, then smart-city projects cannot become sustainable unless their supply chains are environmentally and socially sustainable, or the materials that they need are locally sourced in a sustainable manner.

There are of course considerable challenges to face when the end of a city is inspired by a higher concept, or when the end *is* the higher concept itself. More specifically, building upon Horkheimer's *Eclipse of Reason*, this chapter has accentuated the philosophical and intellectual challenge of defining higher concepts, the planning-related challenge of coordinating all the many activities necessary to realize higher concepts, and the gargantuan challenge of governance since higher concepts comprise a myriad of stakeholders. A city whose end is *sustainability* would then fail if its development was not supported by development in philosophy, urban planning and governance. For Horkheimer (2013), listening to the objective side of reason automatically stimulates philosophical development and, more in general, incites a type of development conducive to higher concepts. Although the German philosopher does not show this process, he elucidates its opposite, and the dynamics of the eclipse or crisis of reason can be therefore used to get a glimpse of what might happen if the phenomenon described by Horkheimer was hypothetically reverted.

In the previous section, the chapter illustrated Horkheimer's crisis of reason, caused by the supremacy of subjective reason and causing 'the complete transformation of the world into a world of means, rather than of ends' (Horkheimer, 2013: 66). Horkheimer (ibid.) argues that subjective reason presses individuals to fulfil personal needs. This stimulus, in turn, translates into technological development, because technology can be directly employed as a medium to satisfy an individual need or sold to generate money (which subsequently becomes a medium itself to satisfy the needs of the individual). Whether in the shape of a practical device or a commodity generating money, technology is, in the eyes of

Horkheimer, a *means* not an *end*. In a capitalist system, like the one critiqued by Horkheimer, it is easy to see how the situation degenerates. This book has remarked that, under capitalism, modernization constantly produces novel technologies, and technological innovation is instrumentalized to provide capitalist companies with new technologies to sell. As part of this technocentric process, *means* multiply and *ends* are eventually forgotten. Assuming that the theory in *Eclipse of Reason* is valid, it is then reasonable to also assume that reversing the processes causing the eclipse explained by Horkheimer might revert the eclipse too. In other words, decreasing the influence of subjective reason could increase the presence of objective reason, thereby preventing the annihilation of higher concepts and ends. However, extreme caution is necessary here. Fully reversing the eclipse depicted by Horkheimer would bring about the opposite scenario, which is an equally problematic prospect: a world of ends, rather than means. The problem would lie in the fact that, as already mentioned, the pursuit of higher concepts requires technology. For example, ends like *sustainability* and *affordable housing* require or, at least, can be facilitated by technological development. A world of only ends is one that does not have the means to accomplish them. If technological development without a vision of development is a blind form of development, its opposite would be a society that clearly sees the destination but does not have legs. It cannot move and reach its destination. Both means and ends are vital and complementary, as well as technology (on the one hand) and philosophy, planning and governance (on the other) are complementary.

Goals are just one component of urban reason. The second end needed by cities is a conclusion. In this regard, an *end* is intended as a point in time *when* or a point in space *where* a phenomenon is considered to be completed and ceases. This book has shown that eco and smart-city projects follow an endless process of urban development. New cities are constantly built and existing cities experience continuous modernization, changing their infrastructure and design to accommodate the latest (not the last) technology. Urban artificial intelligences are part of the same trend. They originate from automated smart technologies and are now consolidating a new breed of autonomous technologies. However, there is no end on the horizon, and no visible point at which technological innovation and thus urbanization are supposed to stop. *New* has an ephemeral temporality and is always replaced by something that becomes newer. The phenomena turning *new* into *old*, modernization and innovation, are rapid and driven by the capitalistic imperative of finding something novel to monetize. Such an imperative attends to the subjective side of reason. It is bereft of objective reason and higher concepts and, because it has no ultimate goal, it will never stop. Urban growth *ad infinitum* is unsustainable, particularly when infinity is not justified by a higher concept. This issue is exemplified by the case of Abu Dhabi where new experimental cities like Masdar City continue to expand in the desert for economic reasons, despite the fact that a community does not exist and people are reluctant to inhabit a non-place. The genesis of the Emirati city and of the many technologies that are there tested is not motivated by any higher concept. Discourses and labels like *sustainability* and *eco* suggest that Abu Dhabi's urban experiment is targeting a goal sensitive to collective social and ecological concerns while, in reality, the Masdarian factory of clean technology is permanently open to protect the economic interests of the local elite and its business partners.

Finding a conclusion and having an objective go hand in hand, since the act of aiming for a target creates momentum driving towards a destination. In urban development, momentum is movement in time and space with cities, for instance, growing additional infrastructures and developing new districts over the years. Resources are mobilized via the extended supply chains of planetary urbanization outlined in Chapter 1. They traverse lands and seas to join the ever-growing and changing city where they are metabolized, thereby expanding the mass of the built environment. Movement is energy. It necessitates a multitude of forces, material and immaterial, ranging from electricity and oil, to labour and capital. Without a conclusion, movement is perpetual and so is the consumption of the energy necessary to sustain it. Cities thus need to stop, but this does not mean that they must terminate their existence. On the contrary, they should continue to grow and develop by following one or more goals inspired by higher concepts, and halt their expansion once all the goals have been realized. New objectives might be added in the future and generate momentum again, but it is crucial that the energy spent moves the city towards a well-defined collective end, instead of feeding an endless modernization whose benefits are unevenly distributed.

On these terms, transhumanism and transurbanism are unsustainable strategies of development because they do not present ends. While they do have a set of objectives, their dimension is narrow. Transhumanism in practise concerns the enhancement of the self via superior technology accessible only to an economically powerful elite. Similarly, transurbanism is about developing advanced urban artificial intelligences that are meant to coexist with riskier and cheaper AI. Collective goals can be present, but in the shape of labels and discourses which cover what is actually happening on the ground. Behind the noble invention of a neurotechnology alleged to preserve the cognitive dominion of mankind over AI, lies a brutal market in which few individuals can afford to upgrade their brain. Behind the honourable perfection of an autonomous car that guarantees the safety of its owner, there is an expensive bill to pay, separating the rich from the poor and the safe from the unsafe. These are the same dynamics of uneven development identified in the empirical case studies, and recurrently denounced in geographical literature (Harvey, 2019; Smith, 2010). Above all, transhumanist and transurbanist cities are endless cities constantly in search of the newest technology. No conclusion is in sight because a given device, whether it is a brain–computer interface or a city brain, can always be upgraded. There is also no interest in finding a conclusion, since the end of innovation would imply the end of the companies and stakeholders whose business and life depend on the commercialization of novel products. The machines of modernization and urbanization are perpetually kept in motion, endlessly burning energy and heading toward no end.

However, even if cities do not have ends, they will nonetheless reach an end. This will be an end of a different kind, compared to the *ends* theorized and discussed above. In the long run, the end of cities with no *ends* is arguably going to reflect one or more of the following three scenarios. The first end is the end of the city as a predominately human space. In this scenario, cities are populated mostly by artificial intelligences such as robots akin to those described in this book, but also (due to the rapid pace of technological innovation) going beyond what is currently imaginable and constructible. Urban activities are carried out largely by robotic entities and the social dimension of the city is drastically altered or, worse, absent when advanced robots can do almost everything that humans currently do. Moreover, AI does not simply populate cities. It plans, shapes and governs them via already existing city brains. This end is thus a conceptual end too, since an environment that is inhabited, planned, shaped and governed by artificial intelligences can be hardly called a city. Although, as noted in Chapter 1, a universal definition and understanding of the city does not exist, regardless of the epistemological stance that is taken, when the term *city* is evoked it is always to refer to a space that is to some extent human. If the *human* gets completely out of the urban equation, it is unquestionably the end of cities.

The second end is the end of the sustainability of cities intended as the culmination of negative irreversible social and environmental changes. This end is the outcome of the dynamics and effects of *Frankenstein urbanism* illustrated in Chapter 6. The findings of this study indicate that contemporary models of urban development tend to produce fragmented settlements that, like Mary Shelley's monster, are made of incongruous components. Fragmentation divides the city socially and separates the built environment from the natural environment. Worst of all, it creates competition among different spaces. Here the city is not a cohesive spatial whole, but rather a patchwork of contrasting and often incompatible elements. Ecosystems are erased to accommodate new buildings, districts and infrastructures whose life is constantly menaced by the advent of economically more profitable urban projects. In this state of tension, AI adds another strain. As Chapter 7 has clarified, a single artificial intelligence does and will not exist in the foreseeable future. In a city where urban planning is not holistic and homogeneous, a plethora of diverse urban artificial intelligences compete against each other, seeking resources, power and space in the attempt to achieve goals that, as the orthogonality thesis posits, might not even be aligned with human values. Cities consume themselves, what is within and around them, and their inner fractures are so deep and wide that they break.

The third and final end is the end of the city at the hands of human beings, as part of a conscious plan of urban destruction. Humans decide to end the city, before the city ends them. This end could be triggered, for example, by the proliferation of autonomous urban artificial intelligences whose development hinders human development. Service robots replacing the human population and city brains running the city according to ideals which threaten or simply penalize human life are not a desirable outcome for human societies. AI, diffused through experimental urban projects pushing the boundaries of innovation from automation to autonomy, turns out to have been a mistake. Urban experiments, now autonomous urban creatures operating with humans out of the loop, also turn out to have been a mistake. Under the belief that it is not too late to save humanity, the creators attempt to terminate their creatures. This end does not necessarily have to be imagined in an apocalyptic fashion. On the contrary, Chapter 7 has stressed that the repercussions of urban AIs can be significantly unsustainable and harmful, even in the absence of a global apocalypse. A privatized autonomous urban transport sector encouraging long commutes and the production of spaces for cars rather than places for people, service robots stealing jobs, and city brains governing cities on the basis of an idea of what is *right* which is *wrong* for many human beings, are rather banal and yet deeply problematic scenarios, and it is therefore plausible to imagine that someone will somewhere try to end the autonomous city or impede its implementation.

This third end mirrors the conclusion of the story narrated by Mary Shelley. The end of *Frankenstein* takes place far away from where the narrative of the book started. In the last part of Shelley's novel, the protagonist, and so the reader, are not surrounded by the books of a library or by the tools of a laboratory. They are surrounded by the ice and snow of the North Pole. Victor has left the literature behind as well as the scientific instruments of experimentation. He now carries a gun, hunting the creature that he himself had brought to life on that dreary night of November. Victor has understood that the experiment has failed, and he is determined to destroy his creation. It is a long chase that pushes the Doctor, not young anymore, towards the edge of the world and the limit of sanity. It is a sad end because Victor dies. The attempt of ending his creature costs his life. However, the so-called monster is not responsible for the final blow. In the end, there is no fight, no war, no sudden apocalyptic event. The drama unfolds slowly, through a series of small but significant repercussions of the experiment. The creature does not kill the creator since it is not evil, contrary to what those who encounter it believe. The very human concepts of *good* and *evil* do not apply to it, due to the fact that it is not human.

What the creature actually is, thinks and feels, Victor will never know, hence the sadness that permeates the ending of *Frankenstein*. Victor could have known and things could have gone differently if he had approached the project in a less individualistic manner, monitoring his creation throughout the experiment. Instead, he designed and conducted the experimental project blinded by his personal needs and desires, listening to the voice of his subjective reason alone. Worst of all, he abandoned his creature, choosing not to follow and guide its development. He voluntarily put himself out of the loop, and therein lies the biggest paradox of Frankenstein's tale. The same paradox that causes the Doctor's demise might cause the demise of human societies in the age of urban artificial intelligences. On the one hand, Victor is a modern human being. He decides to shape his own destiny and embarks on an impossible scientific quest, seeking to build his dream with the instruments of science and technology. He manages to create life, thus achieving his goal and, in so doing, redefining the boundaries of the possible. Because of the way he dreams and acts, Victor epitomizes the concept of modernity introduced in Chapter 1. *Modernity* as the ability to dream of alternative realities and the impulse to realize them. On the other hand, however, once the dream is real, and it is time to protect and cultivate it, Victor gives up its modernity. He stops dreaming, ceasing to be in control of his own destiny and of what he has created. Without even realizing it, Victor turns into a passive agent. The world around him changes and so does his experiment which is not *his* anymore. It is an intelligent and autonomous entity with its own will and an agenda which differs from the Doctor's plan. When Victor wakes up and seeks to regain its modernity it is too late.

The passage from automation to autonomy and the formation of autonomous cities already show one of the two mistakes made by Victor. Eco and smart-city projects as well as urban artificial intelligences are the product of the dreams of individuals and elites, and collective visions of the future are urgently needed to shift urban development away from the sphere of the self and toward higher concepts like justice, democracy and sustainability. Given that AI technologies are largely in the making and the autonomous city is still an emergent phenomenon, Victor's second mistake is not yet present in contemporary cities. As such it is avoidable, but its prevention requires coming to terms with another paradox. If the development of the city and the development of artificial intelligence are meant to proceed together along a track leading to a well-defined and shared end, this would imply dreaming as a collective, not as individuals. Above all, it will be imperative to dream while being awake, nurturing urban experiments and constantly monitoring their behaviour. Dreaming of non-human intelligences and realizing them, making sure that human values are integrated into the city and respected, is about a new modernity in which humans are no longer the only dreamers. Experiments might be dreaming too. In this urban epoch, cities will have to dream with all their eyes wide open.

References

Ahmed, S. (2010). *The promise of happiness*. Duke University Press, Durham.

Aristotle (1996). *Physics*. Penguin Classics, London.

Aristotle (1998). *The metaphysics*. Penguin Classics, London.

Aristotle (2000). *The Politics*. Oxford University Press, Oxford.

Aristotle (2004). *The Nicomachean ethics*. Penguin Classics, London.

Batty, M. (2018). Artificial intelligence and smart cities. *Environment and Planning B*, 15, 3–6. doi:10.1177/2399808317751169.

Bouzarovski, S., Frankowski, J. and Tirado Herrero, S. (2018). Low-carbon gentrification: when climate change encounters residential displacement. *International Journal of Urban and Regional Research*, 42 (5), 845–863.

Bostrom, N. (2005a). A history of transhumanist thought. *Journal of evolution and technology*, 14 (1), 1–25.

Bostrom, N. (2005b). In defense of posthuman dignity. *Bioethics*, 19 (3), 202–214.

Brown, T. (2016). Sustainability as empty signifier: Its rise, fall, and radical potential. *Antipode*, 48 (1), 115–133. doi:690.

Crandall, J. W., Oudah, M., Ishowo-Oloko, F., Abdallah, S., Bonnefon, J. F., Cebrian, M., Sharif, A., Goodrich, M. A. and Rahwan, I. (2018). Cooperating with machines. *Nature communications*, 9 (1), 1–12.

Cugurullo, F. (2013). How to build a sandcastle: An analysis of the genesis and development of Masdar City. *Journal of Urban Technology*, 20 (1), 23–37.

Cugurullo, F. (2020). Urban Artificial Intelligence: From Automation to Autonomy in the Smart City. *Frontiers in Sustainable Cities*, 2, 38.

Eagleton, T. (2014). *Ideology*. Routledge, London.

Ferrando, F. (2013). Posthumanism, transhumanism, antihumanism, metahumanism, and new materialisms. *Existenz*, 8 (2), 26–32.

Foucault, M. (2002). *The Archaeology of Knowledge*. Routledge, London.

Golubchikov, O. and Thornbush, M. (2020). Artificial Intelligence and Robotics in Smart City Strategies and Planned Smart Development. *Smart Cities*, 3 (4), 1133–1144.

Greenfield, A. (2018). *Radical technologies: The design of everyday life*. Verso, London.

Harvey, D. (2019). *Spaces of global capitalism. A theory of uneven geographical development*. Verso, London.

Holton, R. and Boyd, R. (2019). 'Where are the people? What are they doing? Why are they doing it?' (Mindell) Situating artificial intelligence within a socio-technical framework. *Journal of Sociology*, 1440783319873046.

Horkheimer, M. (2013). *Eclipse of Reason*. Bloomsbury, London.

Imrie, R. and Lees, L. (Eds.). (2014). *Sustainable London?: The future of a global city*. Policy Press, Bristol.

Kassens-Noor, E. and Hintze, A. (2020). Cities of the Future? The Potential Impact of Artificial Intelligence. *AI*, 1 (2), 192–197.

Kirkham, R. L. (1992). *Theories of truth: A critical introduction*. MIT Press, Cambridge.

Krueger, R. and Gibbs, D. (Eds.). (2007). *The sustainable development paradox: Urban political economy in the United States and Europe*. Guilford Press, New York and London.

Lagassé, H. D., Alexaki, A., Simhadri, V. L., Katagiri, N. H., Jankowski, W., Sauna, Z. E. and Kimchi-Sarfaty, C. (2017). Recent advances in (therapeutic protein) drug development. *F1000Research*, 6. doi:10.12688/f1000research.9970.1.

McClintock, N. (2018). Cultivating (a) sustainability capital: Urban agriculture, ecogentrification, and the uneven valorization of social reproduction. *Annals of the American Association of Geographers*, 108 (2), 579–590.

Neuralink (2020). Neuralink. [Online] Available: https://www.neuralink.com [Accessed 10 November 2020].

Raco, M. (2005). Sustainable Development, Rolled-out Neoliberalism and Sustainable Communities. *Antipode*, 37 (2), 324–347.

Plato (2007) *The Republic*. Penguin, London.

Smith, N. (2010). *Uneven development: Nature, capital, and the production of space*. University of Georgia Press, Athens.

Swyngedouw, E. (2007). Impossible 'sustainability' and the post-political condition. In J R Krueger and D Gibbs (eds) *The Sustainable Development Paradox*. Guilford, New York, pp. 13–40.

Whitehead, M. (2007). *Spaces of sustainability: Geographical perspectives on the sustainable society*. Routledge, London.

Yigitcanlar, T. and Cugurullo, F. (2020). The Sustainability of Artificial Intelligence: An Urbanistic Viewpoint from the Lens of Smart and Sustainable Cities. *Sustainability*, 12 (20), 8548.

INDEX

Printed in the United States
by Baker & Taylor Publisher Services

Printed in the United States
by Baker & Taylor Publisher Services